OPERATIONAL EFFICIENCY IN FORESTRY

VOL. 1: ANALYSIS

FORESTRY SCIENCES

Baas P, ed: New Perspectives in Wood Anatomy. 1982. ISBN 90-247-2526-7
Prins CFL, ed: Production, Marketing and Use of Finger-Jointed Sawnwood. 1982. ISBN 90-247-2569-0
Oldeman RAA, et al., eds: Tropical Hardwood Utilization: Practice and Prospects. 1982. ISBN 90-247-2581-X
Den Ouden P and Boom BK: Manual of Cultivated Conifers: Hardy in Cold and Warm-Temperate Zone. 1982. ISBN 90-247-2148-2
Bonga JM and Durzan DJ, eds: Tissue Culture in Forestry. 1982. ISBN 90-247-2660-3
Satoo T and Magwick HAI: Forest Biomass. 1982. ISBN 90-247-2710-3
Van Nao T, ed: Forest Fire Prevention and Control. 1982. ISBN 90-247-3050-3
Douglas J: A Re-appraisal of Forestry Development in Developing Countries. 1983. ISBN 90-247-2830-4
Gordon JC and Wheeler CT, eds: Biological Nitrogen Fixation in Forest Ecosystems: Foundations and Applications. 1983. ISBN 90-247-2849-5
Németh MV: The Virus-Mycoplasma and Rikettsia Disease of Fruit Trees. ISBN 90-247-2868-1
Duryea ML and Landis TD, eds: Forest Nursery Manual: Production of Bareroot Seedlings. 1984. ISBN 90-247-2913-0
Hummel FC, ed: Forest Policy: A Contribution to Resource Development. 1984. ISBN 90-247-2883-5
Manion PD, ed: Scleroderris Canker of Conifers. 1984. ISBN 90-247-2912-2
Duryea ML and Brown GN, eds: Seedling Physiology and Reforestation Success. 1984. ISBN 90-247-2949-1
Staaf KAG and Wiksten NA: Tree Harvesting Techniques. 1984. ISBN 90-247-2994-7
Boyd JD: Biophysical Control of Microfibril Orientation in Plant Cell Walls. 1985. ISBN 90-247-3101-1
Findlay WPK, ed: Preservation of Timber in the Tropics. 1985. ISBN 90-247-3112-7
Samset I: Winch and Cable Systems. 1985. ISBN 90-247-3205-0
Leary RA: Interaction Theory in Forest Ecology and Management. 1985. ISBN 90-247-3220-4
Gessel SP: Forest Site and Productivity. 1986. ISBN 90-247-3284-0
Hennessey TC, Dougherty PM, Kossuth SV and Johnson JD, eds: Stress Physiology and Forest Productivity. 1986. ISBN 90-247-3359-6
Shepherd KR: Plantation Silviculture. 1986. ISBN 90-247-3379-0
Sohlberg S and Sokolov VE, eds: Practical Application of Remote Sensing in Forestry. 1986. ISBN 90-247-3392-8
Bonga JM and Durzan DJ, eds: Cell and Tissue Culture in Forestry. Volume 1: General Principles and Biotechnology. 1987. ISBN 90-247-3430-4
Bonga JM and Durzan DJ, eds: Cell and Tissue Culture in Forestry. Volume 2: Specific Principles and Methods: Growth and Developments. 1987. ISBN 90-247-3431-2
Bonga JM and Durzan DJ, eds: Cell and Tissue Culture in Forestry. Volume 3: Case Histories: Gymnosperms, Angiosperms and Palms. 1987. ISBN 90-247-3432-0 (Set Vols. 1–3: ISBN 90-247-3433-9)
Richards EG, ed: Forestry and the Forest Industries: Past and Future. 1987. ISBN 90-247-3592-0
Kossuth SV and Ross SD, eds: Hormonal Control of Tree Growth. 1987. ISBN 90-247-3621-8

Operational Efficiency in Forestry

Vol. 1: Analysis

edited by

U. SUNDBERG
Swedish University of Agricultural Sciences, Garpenberg, Sweden

C.R. SILVERSIDES
Prescott, Ontario, Canada

SPRINGER-SCIENCE+BUSINESS MEDIA, B.V.

Library of Congress Cataloging in Publication Data

Operational efficiency in forestry / U. Sundberg, C.R. Silversides, editors.
p. cm. -- (Forestry sciences ; 29-)
Contents: v. 1. Analysis.
ISBN 978-90-481-8309-8 ISBN 978-94-017-0504-2 (eBook)
DOI 10.1007/978-94-017-0504-2
1. Forestry efficiency. 2. Forests and forestry--Labor productivity. 3. Forestry engineering. I. Sundberg, U. (Ulf) II. Silversides, C. R. III. Series: Forestry sciences ; v. 29, etc.
SD387.E33O64 1988
338.1'6--dc19 88-2940
CIP

ISBN 978-90-481-8309-8

2nd printing 1996

PREFACE

It is our conviction - based on many years of teaching and practical experience - that professional skill in forestry will develop more readily and more efficiently if forestry students are presented with a clear understanding, at an early stage, of the impact of the physical factors that both enhance and inhibit forestry activities. Tnis knowledge is best aquired by adressing the physical world of forestry with a set of analytical tools through which an almost infinite number of variables can be listed in order, be measured and their interaction be explored and resolved. This is the main purpose of this joint effort, presented in two volumes, Part I and Part II.

Part I is analytic, written as a basic text for undergraduates in courses such as logging, transport, forest engineering and even forest management. It deals with the fundamentals of technology in forestry as determined by the physical environment. Briefly it can be visualized as the application of forestry within the parameters: space, time and energy. Forestry is a process with a long time horizon. Free solar energy creates biomass which, with the aid of other forms of energy, is converted into products and services. A knowledge of the dependecies and interactions is indispensible for the forestry profession.

The analytic approach serves two purposes: to bring about a clear understanding of the real world of forest and to develop tools through which efficiency and productivity can be explored and improved.

Reality is somewhat simplified in order to better comprehend the dialectic nature of the phenomena discussed. In Part I large numbers of variables and complex computerized models are omitted because they tend to obscure the basic structures set forth. Complex mathematics and computers will be the next step.

Part II will mainly deal with change and the ways and means to adapt to it. The spectacular increase in industrial productivity was achieved largely by bringing the production factors - labour and capital - together in factories, streamlined and designed for efficiency. Work places in forestry bear no resemblence to these standardized conditions. They constantly change and adaptions will always be necessary and important. Also the time perspective in forestry calls for special professional skill as time - duration and seasonal - in forestry makes special demands which must be mastered.

The two volumes presented should be complementary to the technical books and journals in which the technology of today is described in more detail.

Prescott, Ontario and Krylbo, Sweden 1987

C.R. Silversides and Ulf Sundberg

CONTENTS

1. PROBLEM IDENTIFICATION

1.1 THE PROBLEM

Forests yield goods and services demanded by man. Forestry is concerned with satisfying these demands, to produce these goods and services for the benefit of man. Nature is the basic agent in forestry and creates - through biological processes - the resource - the forests - from which these goods and services derive. To make these goods and services available to man in the right form and at the right place and time requires sacrifices by man - man must, as nature, make an input, a contribution of his own resources. And as mankind increases in number and the forest resource is limited, man's demands tend to successively exhaust nature's own ability to produce. Apart from man's inherent desire to enjoy the benefits of the forest - in quantity and quality - with limited sacrifices, a necessity arises to manipulate the forest resource towards increasing yields through a planned input of human activities. To economize with human activities in forestry is the notion of operational efficiency with which this book is concerned.

It is essential to recognize that the forest resource is a live biological system. Its ability to produce goods and services is or can be strongly influenced by the human activities - it can be improved, retained, decreased or even destroyed. As the forest is both the resource and the product no clear cut distinction can or should be made between the management of the resource and the utilization of its yield.

The best insight into forestry at the outset of our study is a brief look on the revenues and costs of a forest enterprise and the material flow in such a system. We shall choose a big forest company in Sweden, its resources and inputs for the delivery of wood to the mills and for the management of the forest resource. Not all forest enterprises comprise all the phases in this system - some phases are transferred to entrepreneurs or other enterprises - and the size will generally be much smaller. But all activities will have to be performed and it is the efficiency of these activities which are the subject of this book. Independently of who carries out an activity, the cost of it will ultimately have an direct impact on the owner of the forests, on his net profit (stumpage) for wood produced.

The enterprise will be studied under the following headings:

Land assets
Material flow
Costs
Revenues
Operations surplus
Labour inputs and productivity
Silviculture
Machines and machine capital
Profit
Energy
Concluding remarks

All the figures refer to the calender year 1984.

Productive forest is all land producing one green tonne of merchantable stemwood, inclusive bark, per hectare and year over the rotation of a stand. Tonne is the green weight of wood and bark and corresponds approximately to one solid cubic meter of merchantable wood with bark (85% wood, 15% bark).

Table 1.1 The forests of a Swedish company

Area of productive forests	813 000 hectare
Annual increment, per hectare	2,5 tonne
total	2 032 500 tonne
Annual cut, total	1 696 000 tonne
per hectare	2,09 tonne

It can be observed that the increment exceeds the cut with 0,41 tonne per hectare and year which can be looked upon as capital formation in the enterprise.

This enterprise has a fairly even age distribution of the forests and can be regarded as being a type of sustained, long term venture, which has the potential to continue like this infinitely.

1.1.1 Material Flow. The flow of wood is shown in the graph on p. 3. It can first be noted that this enterprise purchased wood from other sources but also delivered wood to other mills than their own industries, which received 3 373 000 tonne in total. The different logging and transport systems are indicated with their respective quantities as well as the share of wood from final cuts and thinnings.

This material flow system is adapted to the forest resource in one end - the beginning - and to the requirements of the mills in the other. In the long term, it is a problem of mutual adaptation which we can regard as part of a forest sector system.

The wood in this case comes from trees - pine, spruce and birch - of fairly small sizes. The average tree may yield around 0,15 tonne of merchantable material and the operation thus involves the logging of around 27 000 000 trees.

WOOD FLOW

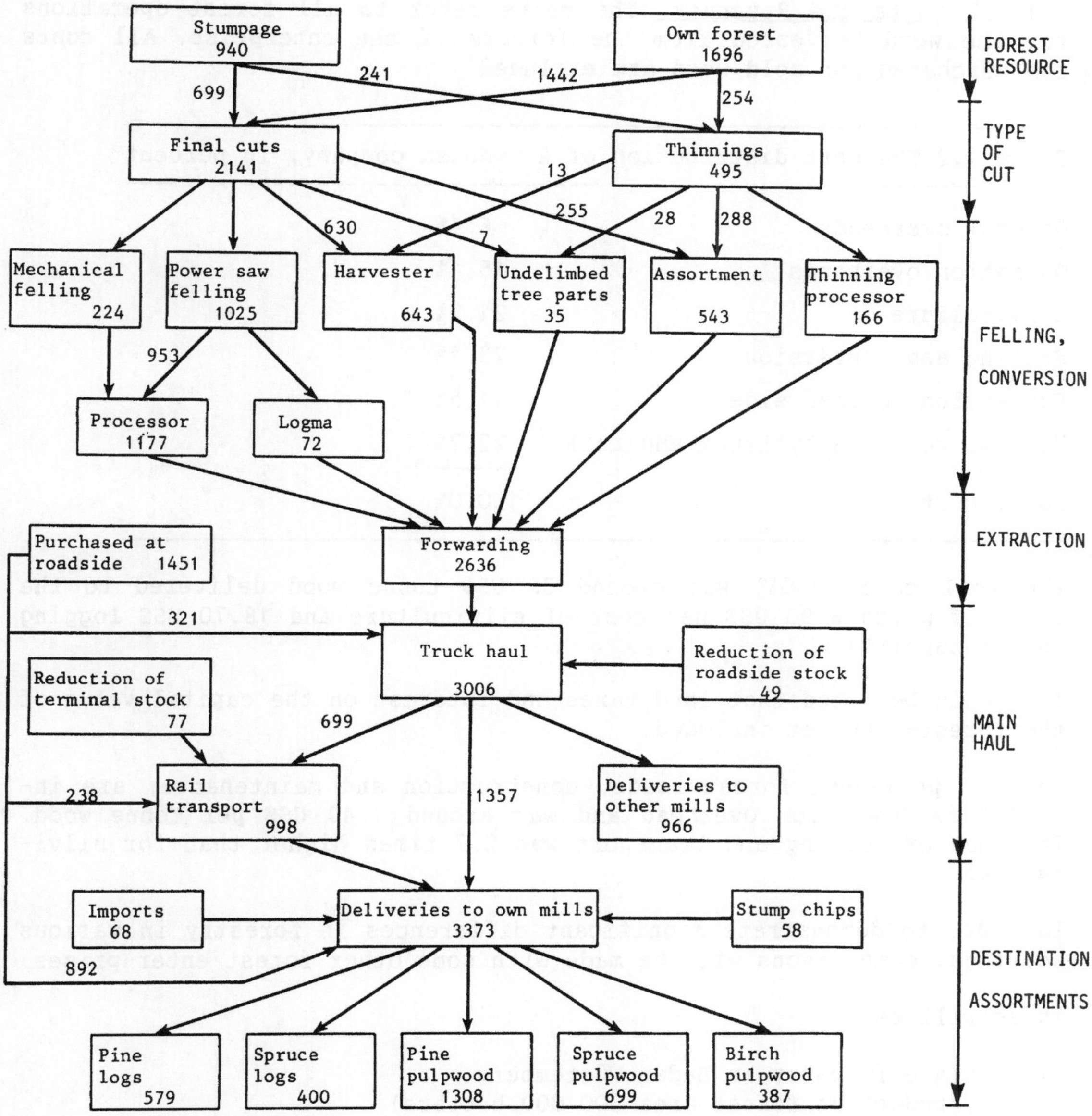

Figure 1.1. Example on the wood flow of a forest enterprise with own forest and mills, also purchasing and delivering wood externally. Stumpage = purchased from other forest owners as standing trees. Numbers in and between boxes = 10^3 tonne

It can easily be visualized that a forest enterprise is a very complex production system, requiring not only skills and competence on all levels but also an information and communication system to cater for changing demands and situations. In particular, the fluctuations in the markets for the processed goods, such as lumber, pulp and paper, will have an immediate impact on the material flow all the way back to the utilization of the standing tree.

1.1.2 Costs and Revenues. The costs refer to all forest operations for the wood harvested from the forests of the enterprise. All costs for purchased and sold wood are excluded.

Table 1.2 The cost distribution of a Swedish company, in percent

General overheads	6.4%
Operation overheads	15.1%
Silviculture	21.2%
Felling and conversion	23.1%
Extraction to road side	11.5%
Haulage to mills by truck and rail	22.7%
Total cost	100.0%

The toal cost (100%) was around 32 US$ tonne wood delivered to the mills of which 6.90 US$ was cost of silviculture and 18.70 US$ logging and transport*).

It should be noted that land taxes and interest on the capital value of the forests are not included.

Cost of permanent forest roads, construction and maintenance, are included in Operation Overhead and was around 1.40 US$ per tonne wood. The cost of logging and transport was 2.7 times higher than for silviculture.

In order to demonstrate significant differences in forestry in various settings, comparisons will be made with some other forest enter-prises.

These will be:

(i) State forestry in Baden-Würtemberg (productive forest area 300 000 hectare)
(ii) A Canadian company

The three enterprises in this comparison are large scale forest operators. But small scale farm forestry has the same cost structure, except possibly for a smaller share of overheads.

---

*) Refer to the year 1984 but money converted with exchange rates of late 1986, 1 US$ = 2 DM = 7 SEK (Sw. Crowns)

Table 1.3 Gross data for some forest enterprises.

	Swedish company A	Baden Würtemberg B	Canadian company C
Growing stock, m^3 per hectare	92	302	225
Annual increment, "-	2,5	7,1	(5,7)
" cut, m^3 per hectare	2,1	7,0	3,3
Revenues, US$ per m^3	46	65	29
" US$ per hectare	96	457	95
Cost, US$ per m^3	32	58	24
" US$ per hectare	68	404	78
Oper. surplus, US$ per m^3	14	7	5,30
"- US$ per hectare	29	50	17,50

A and B are in a steady state, with annual cuts close to the increment. C exploit virgin forests, in which increment is zero. As these areas are harvested the increment will go up and at some point in time balance or exceed the cut. When this will occur will depend on the rate of exploitation. B's higher revenues per m^3 are partly due to proximity to markets, bigger trees yielding more sawlogs and less pulpwood than for A. B's operations are more intensive and costly than A's, partly due to higher inputs for the social and environmental functions of the forests.

Table 1.4 Staff and labour force (conversion factor: 1 m^3 green wood under bark = 1 tonne wood with bark).

	Swedish company A	Baden Würtemberg B
<u>Hectare per man</u>		
Professionals	42 800	1 820
Administrative staff	16 230	2 080
Rangers, technicians	6 800	530
Foremen	10 600	430
Workers	690	120
All categories	560	80
<u>Annual cut, tonne per man</u>		
Professionals	89 260	12 740
Administrative staff	33 920	14 560
Rangers, technicians	14 130	3 710
Foremen	22 030	3 010
Workers	1 440	840
All cathegories	1 180	560

B's inputs of staff and labour is higher than for A both on land unit and unit of production. However, if we compute the revenue per man - the value production per man, the difference will shrink.

Table 1.5 Revenues per employee, US$ per man

	Swedish company A	Baden Würtemberg B	Canadian company C
Revenue, US$ per man	54 000	36 000	175 000x)

x) Estimate 200 workdays per year

A conclusion, if the respective input/output relations for these enterprises are optimal, cannot be made on the basis of these tables. There are a number of factors specific for each enterprise and requirements which do not enter the revenue/cost accounts and do not show up in production statistics. Only few and rather incomplete studies have been made on the complex but important issue on optimal organization of the land/man system that a forest enterprise is. A possible analytical model to address this problem would be the "isoquant", further explained in chapter 4. The enterprises A and B will represents different isoquants as the settings and the end product is different. This is indicated in principle in the figure below.

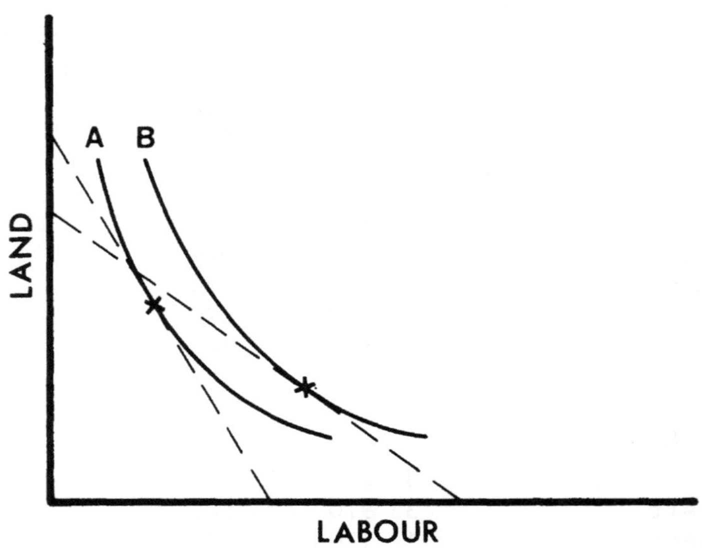

Figure 1.2. Hypothetical isoquants for A and B for the production of one unit (e.g. 1 m^3). Dotted lines are cost lines for land and labour inputs, tangential to the isoquants in the case of optimality.

Table 1.6 Costs and cost distribution on main activities

	Swedish company A	Baden Würtemberg B	Canadian company C
Total cost, US$ per hectare	68	404	78
"- US$ per m^3	32	58	24
Break down, in percent			
General overheads	6	18	1,2
Operation overheads	15		6,4
Silviculture	21	20	12
Felling	23	29	7
Extraction to roadside	12	12	16
Haulage to mills by truck and rail	23	Not included	32
Misc., incl. housing of labour	0	21	26
	100%	100%	100%

Table 1.7 Capital investments - roads and machines - wage rates

	Swedish company A	Baden Würtemberg B	Canadian company C
Forest road, meter per hectare	12	48,6	NA
Machine investm*), US$ per hectare	42	46	22
(Replacement value), US$ per tonne	20	6,60	3,64
US$ per worker	29 000	5 520	72 000
Wage rate for workers, US$ per hour	6,30	7,40	13,00
Fringe benefits, percent	60	123,8	20
Total labour cost, US$ per hour	10,10	16,60	15,60

*) Logging trucks and vehicles for staff and labour not included.

The output in felling and extraction for delivery at roadside depends very much on the tree size and examples are shown in table 1.8.

Table 1.8 Output for felling and extraction to roadside

	Tree size, m^3	Tonne per man day
Swedish company	0,15	13,0
Baden Würtemberg	0,25	11,6
Canadian company	0,28	42,4
New Zeeland, radiata forests		
tractor logging (incl. loading)	1,5	31-55
relogging	NA	15-20
road line salvage	1,5	19
highlead	1,5	23
skyline	1,5	19
thinning, shortwood	0,1-0,2	7-8

Note: In Table 1.8 figures from a forest company in New Zealand are included as comparison, showing a wide variation for logging methods and tree size.

The historical development of mechanization in Sweden is shown in the graph, Figure 1.3.

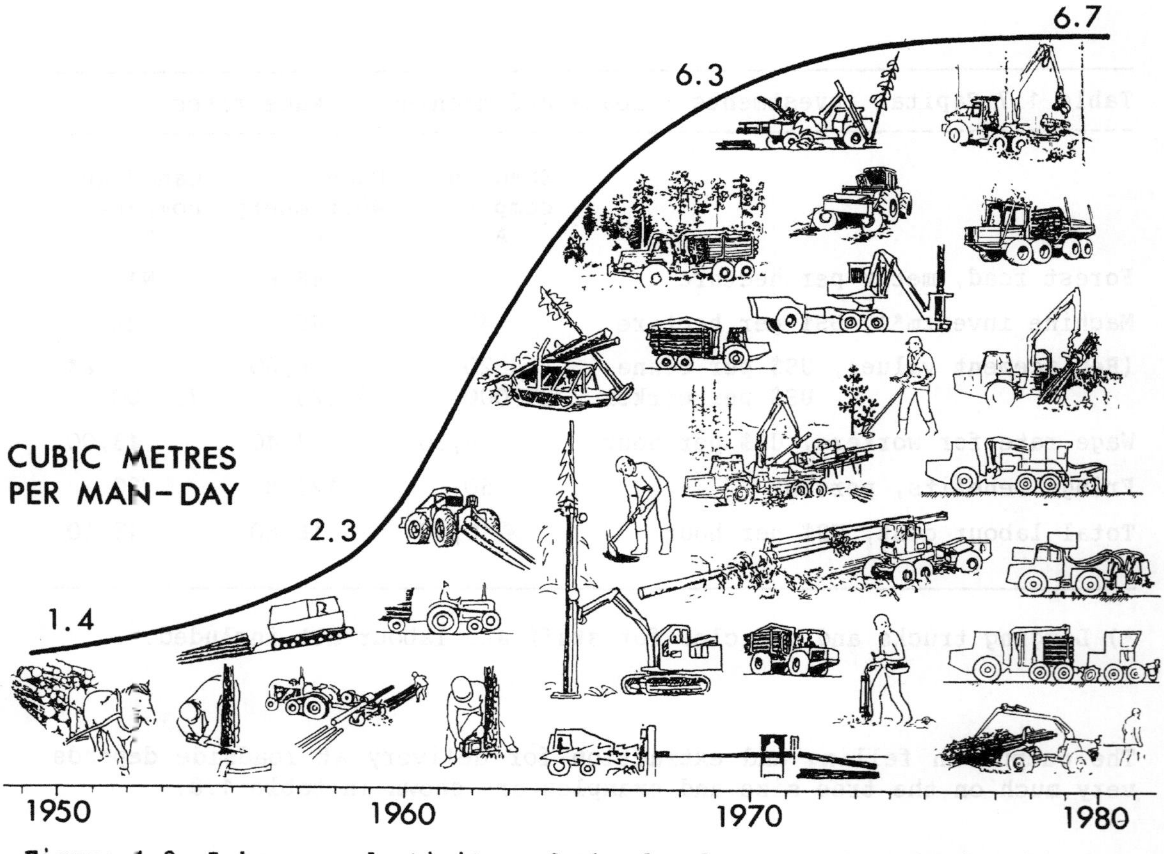

Figure 1.3. Labour productivity and the development of mechanization in Swedish forestry. Courtesy: Stig Andersson

1.1.3 Energy. Based on other studies (e.g. Sundberg, 1982) the enterprise A is shown as an energy system in the figure below, expressed in joule per hectare.

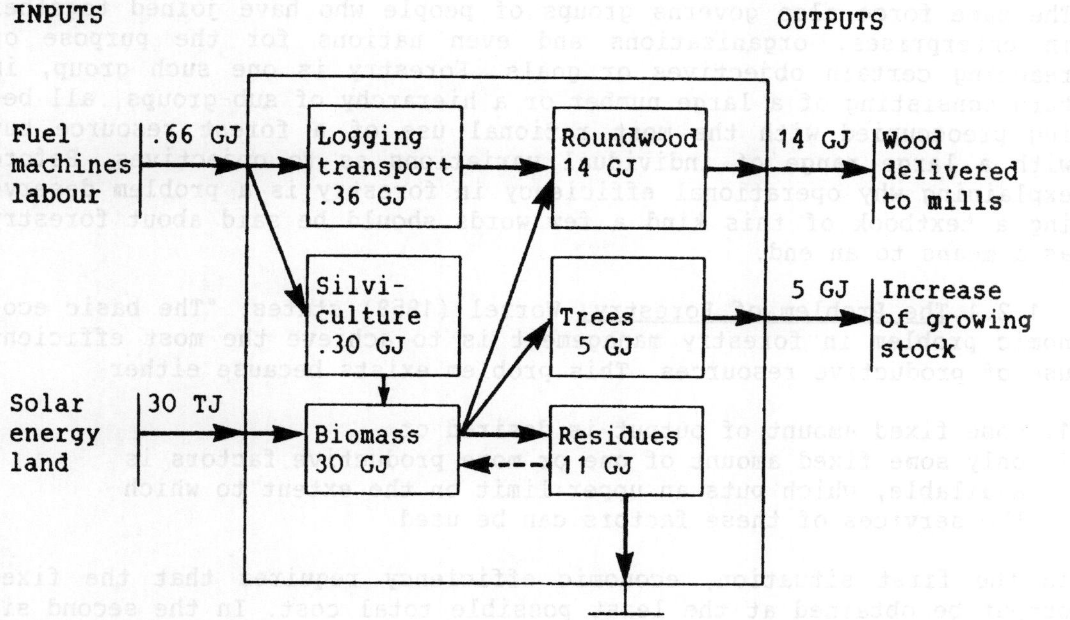

Figure 1.4. The material flow in figure 1.1 (p. 3) for the Swedish forest enterprise A as an energy system. Per hectare forest land. TJ = terajoule = 10^{12} joule; GJ = gigajoule = 10^{9} joule.

Disregarding the solar energy, usually considered as free energy, the input/output ratio is around 1:20 (0,66:14) or inputs are around 5% of the output. The inputs include direct energy as fuel, indirect energy as embodied in the machines and energy for transport and food for labour. Looking on forestry as an energy system its efficiency is very high, provided solar energy is free.

1.1.4 Concluding Remarks. This study of the structure, operation and economics of forest enterprises is presented to give an understanding of the skills and knowledge required to plan, supervise and control a forest system. Such a system is subject to changes imposed from the outside. It is the main function of a forester to respond to such changes and reorient the operations efficiently in conformity with the goals of the system or enterprise.

It should be noted that the studied Swedish enterprise operates in a situation with an excellent infrastructure of roads, railways, housing and social services. In developing countries or in remote regions forestry may have to carry substantial parts of the cost for such facilities.

1.2 WHY IS IT A PROBLEM

All problems in practical life derive from conflicting demands. Such problems are present in each individual when he or she wants to achieve much with a small effort. This inherent quality of man - his laziness - is often assumed to be the basic force for invention and development. The same force also governs groups of people who have joined together in enterprises, organizations and even nations for the purpose of reaching certain objectives or goals. Forestry is one such group, in turn consisting of a large number or a hierarchy of sub-groups, all being preoccupied with the most rational use of a forest resource but with a large range of individual variations as to objectives. Before explaining why operational efficiency in forestry is a problem deserving a textbook of this kind a few words should be said about forestry as a means to an end.

1.2.1 The Problem of Forestry. Worrel (1959) states: "The basic economic problem in forestry management is to achieve the most efficient use of productive resources. This problem exists because either

1. some fixed amount of output is desired or
2. only some fixed amount of one or more productive factors is available, which puts an upper limit on the extent to which the services of these factors can be used

In the first situation, economic efficiency requires that the fixed output be obtained at the least possible total cost. In the second situation, economic efficiency requires that the factors be used in such a way as to produce a maximum net return to the limited factor."

This distinction is fundamental for the understanding of forestry operations at large, in particular with regard to recognizing the difference between the two classical forms of forestry - exploitive and sustained yield forestry. This difference is not that exploitive forestry deals with the final harvest of virgin or mature stands and that sustained yield forestry works with forests of all ages. The difference is that in sustained yield forestry, the limiting production factor is the forest resource whereas in exploitive forestry the forest resource - for all practical purposes - is unlimited and shortage exists in some other aspects. The latter shortage can be such things as:

markets - a limited demand for forest products or limited possibilities to process or market these products, whatever they are

capital - limiting the possibilities to expand

labour - limiting the possibilities to produce

In exploitive forestry one such shortage factor often dominates at a certain time or at a certain stage and is then often superseded by another. When all of them are in sufficient supply forestry in that particular location passes over from exploitive to sustained yield and the forest resource has become the shortage factor. Viewing forestry developments from this angle it is evident that forestry in all regions has started as exploitive and that forestry everywhere in due course is bound to end up as sustained yield forestry. But this time is not yet here.

The distinction between exploitive and sustained yield forestry has been somewhat elaborated here because of their quite different objectives and different criteria on efficiency, and these differences tend to bear rather strongly on ways and means for organizing and executing the forestry operations. Consequently, the distinction is essential for evaluation or assessment of the efficiency of these operations. (As shall be further explained in the Chapter 2, situations of the same structure show up when it comes to work efficiency problems.)

Exploitive forestry - by definition in this text - has the objective to maximize the profit per unit of production. As already explained this objective is relevant when the forest resource is unlimited. When this resource is in shortage - sustained yield forestry - the objective is to maximize the profit per uit area of forests. Examples can readily be given. Enterprises exploiting tropical forests as concessionaires often have large concession areas with a production potential far exceeding the established capacity for logging, sawmilling etc. Their objective then is to maximize the profit on the production this capacity can give, which is the same as maximizing profit per unit of production. Their interest or incentive to use secondary species or small sized wood is non-existent as the harvesting of such products reduces their production of the prime logs on which their profit is higher. If possibilities exist to increase this capacity it would still be more profitable to increase the productions of prime logs rather than to use secondary species as long as the forest resource from their point of view is unlimited. This view is often very short term because of all kinds of uncertainties in such ventures.

Forestry in Western Europe has since long passed the exploitive period. Markets, labour and capital are at hand and its objective is to maximize the profit per unit area. Each tree or each log with a value exceeding the direct cost of harvesting is utilized and all efforts are directed towards increased production. European foresters visiting other continents are often amazed, sometimes shocked, when they see the wastage of wood in some logging operations. However this wastage has a rational basis and needs not necessarily be wrong from whatever basis the practices are judged. It is the consequence of two different situations and the situations for forestry vary infinitely. This variation which is present both in space and time accounts for the problems of forestry, which are to combine and make use of the available resources to meet the objectives of forestry - to satisfy human demands in the best possible way.

1.2.2 <u>The Problem of Operational Efficiency in Forestry.</u> It is evident without much afterthought that operational efficiency - to economize with human or man-made inputs - is influenced by the higher level objectives of forestry as briefly discussed in the previous paragraph. Already the choice of the right technology for each such situation is a problem requiring knowledge and skill to be correctly solved.

But once this scrutiny in order to adapt the work operations to over riding goals has been made, there remain problems of planning, implementing and execution of these work operations - the problem to allocate in space and time labour and machines, to put them to work in a rational fashion and to maintain or improve their efficiency. Such problems can be grouped in three main categories - social, technical and economic.

The social issues in this context refer to labour and can briefly be specified as follows:

- to provide safe work places for the labour force
- to minimize health hazards
- to design the work so that it is adapted to the capabilities of the workers
- to design the work so that the workers get satisfaction of their work, not only in the form of earnings but also in values other than money.

All these issues are related to the working and living conditions of the labour force. Social issues related to people outside the forestry system such as recreation belong to overriding forest policy problems which are outside the scope of this book.

The technical issues are related to the input of man-made resources in the work operations such as the use of tools and machines, the forest roads and other routes for transport and communication as well as other fixed installations for facilitating the operations. It is the problem of selecting, maintaining and improving the technical means which can perform the job.

The economic issues are to balance the inputs of man, machines and other assets needed for performing the job so as to meet the objectives. Economic objectives may often be complex but it is generally a matter of performing the work to a satisfactorily low cost.

In the early history of forestry, operational efficiency did not deserve much attention in the sense that it was subject to systematic study or teaching. The art was learned in an empirical way. But much has changed. In the first place it has been learned that the efficiency of work can be greatly improved. Secondly technology has brought about technical alternatives by which a job can be performed whereas formerly often only one, or possibly a few, techniques existed from which a choice was rather simple. Thirdly, forestry research has for some decades included the systematic study of work operations through which now so much knowledge and insight has been collected as to explain not only what happened in a work operation but also why it happened. This has opened up the possibilities for predicting with a reasonable degree of certainty the results of various courses of action. Fontaine (1971) expressed this radical change with regard to scientific management in forestry as follows:

> "The role of science and technology is reversed. In the past technology preceeded science, while now scientific discovery inspires technology."

Even if Fontaine in the first place was thinking on silviculture, his statement nevertheless holds true also for work operations in forestry. The practical implication of this change is that the output and costs for carrying out a job with alternate methods and machines can now be predicted. One does not have to carry out costly trials. Available machine systems can now be allocated to work best suited to their respective technical abilities. In other words, forest work operations can be better planned and human and machine inputs optimized.

1.3 WHAT DO WE NEED TO KNOW - THE SHOPPING LIST - MEASUREMENT

What we need to know for solving problems of operational efficiency is:

Which are the technical means for doing the job
What can labour perform equipped with these means
Which factors have influence on the performance, and how
What are the cost of the inputs
What are the prices on the outputs
What restrictions are present in doing the job

Our knowledge needs to be quantitative - expressed in figures. It must be systematic in the sense that it is coherent, consistent and verifiable. And finally, it must be usable so that it can be understood and accepted by the cadre of planners, managers and supervisors of forest work operations - or this cadre should be familiarized and trained to use this knowledge.

These are the main items on the shopping list of knowledge. The list contains a large set of sub-items. In setting out it must be clearly understood that all these desired items are not and will never be available on the supermarket of knowledge. Some items will be missing, many will be available but in an imperfect form, unsatisfactory for our needs. Some of the missing items can be identified and processed, whereas others will be too expensive to "manufacture" simply because the resources to make them available are limited. And - finally - there are items which are unknown, factors influencing the work processes that nobody has yet identified.

The shopping list suffers from all kind of imperfections but this is nothing unique. This is the case in all decision making. To make a wise decision with imperfect knowledge is precisely the true distinction of a good manager but it is also one of his qualities to use the knowledge that is in fact available.

Another feature of the items on the shopping list is obsolescence. Quantifications are often historical data whereas problems as a rule are focussed on the future. There is a time lag, a gap, which can never be completely bridged and which will add to the uncertainties of our conclusions. But these uncertainties can partly be reduced by putting the knowledge in the context of change, by attempts to elucidate what changes are likely to occur and how they will influence our problem in general and its components in detail.

This chapter will deal with details of the shopping list - what are important items and how to measure them. Some will be dealt with rather briefly, in particular these considered less important as well as these easily available elsewhere in literature. The selection and the depth of penetration is arbitrary and reflects merely the authors' assessment of what at present is most fruitful in the education of forestry students and staff.

1.3.1 Production Forms. Forestry is often subdivided into three main phases of production:

primary production - or the growing of trees
secondary production - or the harvesting and
tertiary production - or the processing

This division is somewhat arbitrary and refers primarily to the main item of the product mix in forestry, the production and processing of wood. The third phase, the tertiary production, can for some products be broken up in several phases. Its first step, production of lumber, pulp, paper, board, veneer and plywood is in the USA termed "primary manufacturing". Marketing is also a function which sometimes may deserve recognition as a separate phase.

There are many ways in which this production is organized. In cases when the whole chain of production from the seed to the ultimate consumer of the forest product is in the hand of the same enterprise, the production is fully vertically integrated. However, such complete vertical integration rarely occurs. Most often the production is broken up into entrepreneurial phases, and for each phase some type of enterprise is organized to carry out the various jobs within that phase. This subdivision has several repercussions of importance not only for the efficiency of the whole forestry process but also on the way the work operations are carried out. It has implications on the labour management and employment policies of the enterprises, on the intensity of silviculture, on the incentive to mechanize and invest etc. Consequently, we need to know the inherent characteristics of the forestry production system, the forces and incentives governing the work practices of the enterprises within this system as these forces determine the management policies of the enterprises for the resources under their responsibility such as forests, labour, machines etc. One of the reasons is that each enterprise has its own objective which does not necessarily conform with either the system objective or those of the other enterprises.

Examples on some production forms illustrating various ways and degrees of vertical integration are outlined in Figure 1.5.

Each line denotes an entrepreneurial organisation and its involvment in the work operations. Dotted lines mark options. - Marketing takes place in each node.

The four alternate production forms in figure 1.5 are arbitrarily chosen examples. In real life there are an almost infinite number of such forms and one may very well ask why it is so. Firstly, the reason may be a historical one and existing forms may just be relicts from times when production forestry started, without reflecting any adaptation to existing situations. The resistance to change of the production forms is often extremely great because of the vested interests of the enterprises and groups of people. Secondly, the situations for forestry vary vastly and a production form which is rational in one situation is not necessary the best in another one. Thirdly - and this is perhaps the most important factor - enterprises have often limited possibilities to encompass all the phases or work operations they consider desirable and consistent with their objective. Such limitations can be of different kinds. The forest resource might be owned or controlled by somebody not willing to give it up. This is often the case with publicly owned forests. The enterprise may not be able to raise the capital needed for integrating a new phase in its activity. Or it might not have the competence and skills needed for the new work operations and to build up this competence might be too costly, time consuming or otherwise difficult.

Historically, the owner of the forest resource was in a position to decide or strongly influence the production form, provided he was in full control of the resource. This position still applies when new untouched forest resources are being utilized. But for some practical reason his choice may be limited. His resource may be too small to support the size of enterprise needed or he may not be able to raise the capital. The political or economic structure of the country may make some options politically unfeasible. Private interests in capitalistic countries strongly object that state owned forest industries are developed, based on state owned forests, whereas in countries with planned economies with the forest owned by the state, private forest industries are not acceptable.

Let us return to the figure 1.5 for a closer scrutiny of the four examples on production forms. Alternative I consists of a chain of four enterprises, E_{11} to E_{14} from the growing to the final use of wood. It will be a very rare coincident if all these four enterprises have not only conforming objectives but also that they agree with the objective or goal of the whole production system. Sub-optimization may easily occur because of conflicting objectives or imperfections in the "nodes" between the enterprises where the product is transferred from one to another. A perfect market - a pure competition - is the first prerequisite for this alternative to function well and as soon as some form of monopoly exists the production form is easily thrown out of balance. If monopoly or oligopoly exists in one of these entrepreneurial links, e.g. one or a few firms in the link E_{13} process the wood, these firms will be what in economics is called monopsonists in the roundwood market and try to depress the roundwood prices below their appropriate level. This in turn will bear strongly on the work situations for the preceeding enterprises E_{11} and E_{12}. The firm in E_{13} will similarily be oligopolyist towards the firms in link E_{14} and will tend to establish price levels meeting his particular objectives.

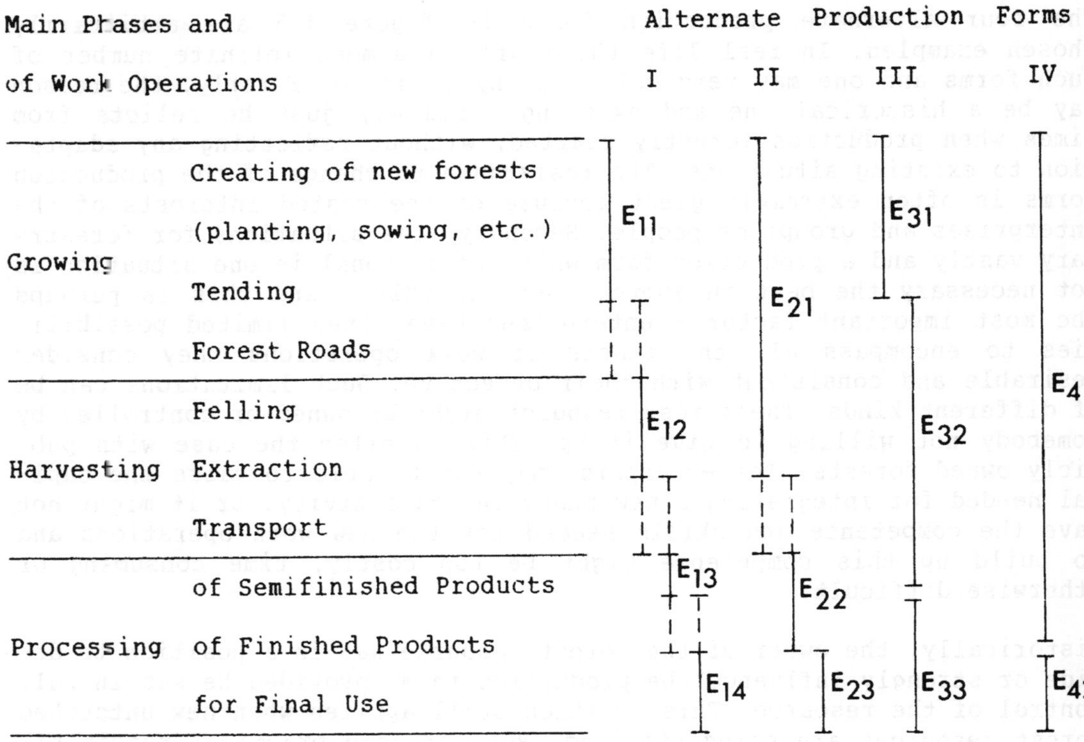

Figure 1.5. Examples on production forms and vertical integration in forestry. E = enterprise.

In Alternative II in figure 1.5 the growing and the harvesting is integrated into one enterprise E_{21} and provisions should exist for an optimum coordination of these activities. The roundwood is than sold to the processing industries in link E_{22}. If there is a true competition for roundwood this system should function well but if there are only one or a few buyers this may not be the case. One means to counteract the latter situation is the formation of "Forest Farmers Associations" with the purpose of negotiating the terms of roundwood sales on a more equal footing - the farmers establish a monopoly in bargaining in the monopsonistic market place.

In Alternative III in figure 1.5 one enterprise E_{32} does the harvesting and processing of semifinished products. This is a very common production form in countries with excessive forest resources, e.g. Canada, USSR (to some extent) and in many developing countries. The enterprise often obtains a concession (in Canada called "limit") to utilize the timber crop from a specified forest area over a certain period of time, often 20-50 years or more. The silviculture often remains with the land owner, the enterprise E_{31}. The terms of the enterprise for using the timber crop are laid down in a contract which may or may not foresee an periodic adjustment of the royalties or stumpage fees to be paid on harvested timber or of the rent per unit area of forest land within the concession. Some of the reasons why this production form is very common are repeated:

(i) The forest owner may not be able to raise the capital needed for starting up logging and processing operation.

(ii) he lacks the competence.

(iii) the marketing of the products from the forest industries may involve great or insurmountable difficulties.

(iv) the political or economic policy of the country may in the case of state owned forest prohibit a vertical integration.

(v) concessions are a perfect - perhaps the best - arrangement to conserve or establish corruption.

In the mind of the authors of this book the production form with concession is in the majority of cases the least efficient of all. Apart from its susceptibility to corruption it suffers from some strongly conflicting objectives of the forest owner and the concessionaire and even if the concession contract is written to solve these, these conflicts are in reality seldom bridged in a satisfactory manner.

Alternative IV marks the most vertically integrated production form. Such integration tends to occur as forestry matures in a country. In an enterprise of type E_{41} the same imperfections may exist as in other less integrated production forms due to problems of information or communication between its different departments or to other administrative difficulties. But in principle, it should be the one best equipped to meet the objectives of forestry both in the short and long term.

It should be observed that many production forms may exist side by side in the same country and also that different products such as sawlogs and pulpwood, lumber and pulp may be produced somewhat differently by the same enterprises. One forest owner may have a sawmill utilizing his sawlogs but market the pulpwood etc.

Vertical integration contradicts one of the fundamental principles for progress and increased efficiency - specialization. And with specialization often follows <u>horizontal integration.</u> The fact that horizontal integration sometimes has greater advantages than vertical is the explanation for the many entrepreneurial phases from raw material to finish products. Horizontal integration offers greater opportunities to exploit and develop both skills and products and may sometimes facilitate more efficient use of equipment. Truck hauling of roundwood may often be more efficient and cheap if carried out by a specialized trucking enterprise, which may move not only logs but also other goods in the same locality. Such a trucking enterprise may be able to arrange the maintainance of the trucks more cheaply and also keep them running for longer periods as it can draw on demands on transport services from many customers. However, the merits of horizontal integration are most likely more apparent in forest industries and related marketing functions than in the growing and harvesting of wood.

This rather lengthy discussion of production forms is made for a clear understanding of the framework within which forest work operations take place. These forms determine the objectives of the separate enterprises in the forestry system and these objectives bear directly on ways and means these enterprises choose for carrying out their activities. Existing production forms are generally based on rationality but not always so. They may sometimes be a heritage from times bygone and retained because of resistance to change in existing structures. In the planning of new forest ventures the question of production form deserves careful analysis of the potential and constraints of the alternatives. For such an analysis the science of economics is a useful tool.

1.3.2 <u>Ownership, Tenure of Land.</u> The possibilities to raise capital has a strong influence on the forms of production. In forestry such possibilities are closely related to the ownership of forest and forest land.

The important thing with real property is that it can be used as collateral against a loan. Real property can be mortgaged and mortgages provide a safe security for loans. The possibility for loans, with real property as security is perhaps one of the most important institutions of the West which facilitated the increased productivity in agriculture and forestry through which the West is able to feed so many people adequately.

In the western culture private ownership of land was introduced and established as an institution at an early stage of the histories of these countries. In North America this is particularly true of agricultural land. Less so with regard to forest land. In Canada for example 90 percent of the forest land is owned by the state, in the United States 21 percent is owned by one or more levels of government while 72 percent is owned privately.

In Canada it is possible for forest industries to obtain long term or "evergreen" rights to state owned forest areas which guaranties them supplies of timber and a degree of permanency comparable to freehold ownership. In the United States forest industries are able to obtain short term rights from state owned forests. Privately owned forests tend to be smaller in area than those owned by the state. Many forests are owned by investment agencies such as insurance companies or pension funds and are leased on long terms i.e. 99 years to forest industries.

The forests in socialistic countries and in many developing countries are owned totally by the state. In such countries the state must provide the capital for all forestry operations or give guarantees to those who exploit the forests so that they in turn will be able to attract the needed capital for their ventures. One can readily understand why the growing of trees and investments in silviculture is largely neglected in many countries with collective ownership of the land. This is sometimes referred to as the tragedy of the commons. What belongs to everybody belongs to no one. Such countries often have many more urgent needs and demands for the immediate support of people and the often long term investments in forestry receive very low priority.

1.4 LOGGING SYSTEMS

1.4.1 General. The harvesting of wood is usually subdivided into three main phases: cutting, extraction and long-distance transport. Within each of these phases there are a number of sub-operations such as the conversion of the felled trees into logs, the moving of the wood and its loading and unloading. Two features of the harvest are note-worthy: firstly, the various sub-operations are interdependent in the sense that an earlier operation influences a subsequent one, and similarily, a later operation may require some preceeding jobs to be done in a particular way. Secondly, several technical options exist for each work element with regard to method of work and input of labour and machines.

The scope of logging is to deliver the wood in the desired form to the consumer - the processing mill - either at the lowest possible cost (the scope of the consumer) or at a cost not exceeding the buyer's price (the scope of the forest owner). The use of the term "logging system" implies that the aggregate cost of harvesting shall meet one of these or other similar criteria. The concept emphasizes the goal of designing a combination of jobs or sub-operations meeting the scope of low cost and not considering each sub-operation in isolation.

Logging systems are often termed according to either the form in which the wood is transported or the method and equipment used, or both. The main systems are thus a combination of modes of work of the following three features:

(a) Conversion of trees to logs

whole tree logging
tree length logging
short wood logging

(b) Extraction methods

forwarding with animals or tractors
skidding with animals or tractors
extraction with cable equipment

(c) Long-distance transport

with trucks
on or in water
by rail

Figure 1.6 illustrates the flow of timber in the most common harvesting systems. Some broad indications on the feasibility of methods and equipment follow as influenced by physical factors. In addition, other external conditions may bear on the choice, such as the pattern of the roundwood market for the different assortments, vehicle specifications for truck hauling on public highways and so on.

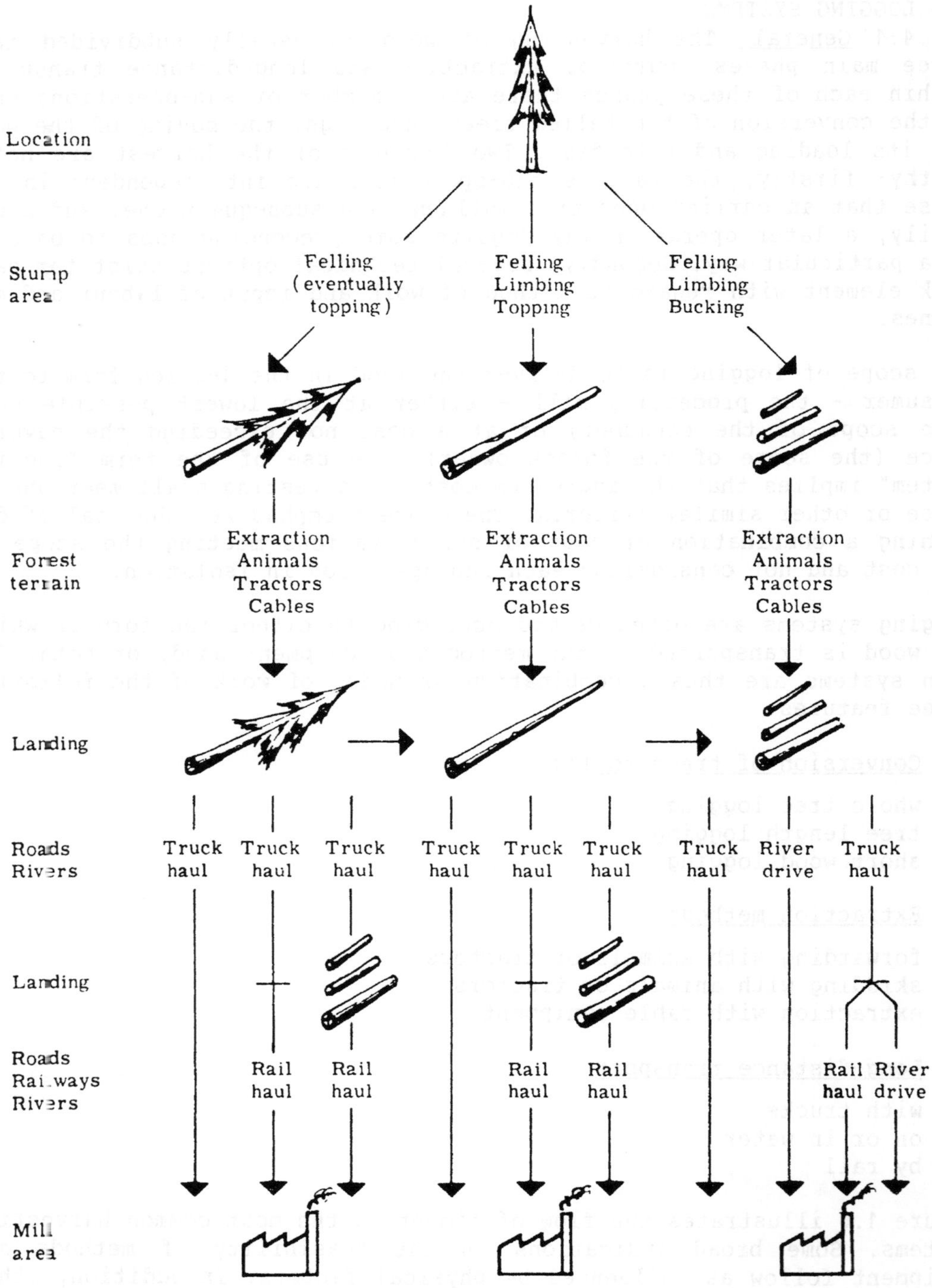

Figure 1.6. Main logging systems

1.4.2 <u>The Choice of Extraction Methods and Long Distance Transport.</u> There are a number of alternative ways of combining methods and equipment. However, an experienced logger can usually narrow down the choice to a few. There are always some features in each situation making options unfeasible.

The following scheme for the choice of off-road transport methods in temperate forests is an attempt to summarize the kinds of terrain and topography in which the various modes of extraction methods are technically feasible.

Table 1.9 Scheme for the choice of off-road transport methods in temperate forests

TERRAIN AND SOIL		GROUND SKIDDING METHODS							CABLE SYSTEMS		
Ground roughness	Soil strength	Manual	Animals	Farm-tractors	Skidder	Forwarder	Crawlers	Feller buncher	Short <700 m Mobile	Longer distances >700 m All-terr.	Gravity only
STEEPNESS	0-10%										
Smooth-	good	x)carry	xx	xx	xx	xx	Swamp	xx			Only if animal and tractor are impracticable
rough	bad	x)drag	x	x	xx	xx	wet clay}xx	xx	xx	xx	
Very	good	x)roll	x		xx	xx		xx	xx	xx	
rough	bad	x)			xx	x	" x	x	xx	xx	
STEEPNESS	10-50%										
Smooth-	good ↑	<40>40%	x<15%	xx<15%	xx<25%	xx<25%	x<35%	xx<25%			
rough	" ↓	x xxgravity	xx<30%	xx<25%	xx	xx<45%	wet	xx<35%		<30>30%	
	bad ↑				x<20%		loamy x<25%	x<20%	xx	xx x	xx <30%
	" ↓	x x gravity	x<30%	x<25%	xx	x<45%	xx	xx<35%	xx	xx x	xx "
Very	good ↑				x<25%	x<25%	x<35%	x<25%	xx	xx x	xx "
rough	" ↓	x gravity	x<30%	x<20%	xx	xx<45%	x	xx<35%	xx	xx x	xx "
	bad ↑				x<20%		" x<25%	x<20%	xx	xx x	xx "
	" ↓				x	x<45%	x	xx<35%	xx	xx x	xx "
STEEPNESS	50% +					Flexible					
Smooth-	good ↑					track			xx	x	xx
rough	" ↓	xxgravity			xx<60%	x<60%	x<60%		xx	x	xx
	bad ↑								xx	x	xx
	" ↓	x gravity				x<60%	" x<60%		xx	x	xx
Very	good ↑								xx	x	xx
rough	" ↓	x gravity			xx<60%	x<60%	x<60%		xx	x	xx
	bad ↑								xx	x	xx
	" ↓					x<60%	" x<60%		xx	x	xx

Legend: ↑ Uphill ↓ Downhill xx Suitable x Possible
Slope in % disregard slopes shorter than 50 m.

Note: The scheme considers technical features only. Final choice based on economics and type of cut.

Comments on the scheme Table 1.9

(A) Topography, terrain and soil

The topography in a plantation area affects the logging method and its productivity in the same manner as in natural forests. On steeper slopes the felling and conversion phase of the operation is less productive and more costly. Short-distance ground transportation must be made straight up or down the slope unless strip roads are bulldozed and feeder roads must be spaced more closely because of the difficulties in skidding from both sides unless one of several possible cable yarding systems is used.

In the classification of topography, terrain and soil the IUFRO[1]) proposal for an international system of terrain classification is recommended. The following points should be observed:

a) Steepness refers to major slopes, longer than 50 m. Slopes shorter than 50 m should be disregarded;

b) Ground roughness refers to the occurrence of obstacles of more than 50 cm height or depth. The classification in "smooth-rough" and "very rough" is somewhat arbitrary but in "very rough", the average average distance between obstacles should be less than 3 m;

c) Soil strength (the resistance against the impact of forces) cannot be quantified in a simple manner. In general "good" applies to friction soils and "bad" to cohesion soils. If possible, classification should be based on practical local experience from the use of tractors, taking into account the form and quantity of precipitation, season of operation and so forth.

(B) Ground skidding method

Ground extraction methods are more economical than cable yarding systems and should be used in most situations when conditions permit. Animals, both horses, mules and oxen, may be used to skid logs on downgrades of 30-35% provided the ground is reasonably smooth and on adverse grades up to 15-20% although loads are considerably reduced on such grades.

Skidding and forwarding machines can operate successfully on downgrades of up to 30-50% depending on vehicle specifications and on ground roughness. Skidding machines are severely handicapped on adverse grades, maximum payload volumes dropping 2.5% for each 1% of grade. Forwarding machines - which carry their payload clear of the ground - can negotiate steeper adverse grades, given sufficient power and good traction conditions:

---

1) IUFRO = International Union of Forest Research Organizations

a) farm tractor equipped with winch and when needed antiskid devices (chains, half-tracks). Auxiliary equipment, skidding tongue, pan, trailer, sled and crane as suitable;

b) articulated, all-wheel-drive skidder, one- or two-drum winch, often small bulldozer blade and supports for winching;

c) forwarder, articulated, all-wheel-drive, or with flexible tracks, winch, crane, often small bulldozer blade.

(C) Cable systems

Cable systems suffer more from low density of cut and from small or very large tree sizes than ground skidding methods. High lead yarding is normally done uphill, requiring the building of feeder roads on the ridges or along the contours. (Ground skidding on slopes, on the other hand, being handicapped severely on upgrades over 10%, is preferably done downhill, requiring that truck roads be located in the valleys.) Feeder truck road spacing is limited by the fixed reach of single-span cable yarding systems.

Steep slopes may be logged using light double-drum winch and cable crane systems. Although these systems perform better when skidding is uphill, they may be used downhill as well. When plantations are thinned with such systems great care must be exercised to prevent damage to the residual stand. It is imperative that parallel installations be used rather than the fan-shaped one working from a central point.

A high lead system may be used with a tractor-mounted double-drum winch where the maximum skidding distance to the truck road or skidder trail is less than 100 m and a cable crane system when skidding distance ranges between 100 and 400 m. Where the skidding distance exceeds 400 m, either an intermediate truck road must be built so that most of the logs may be winched to the roadside, or a tractor-mounted cable crane may be considered. Lightweight cable cranes are limited physically to a yarding distance of 800-900 m, so that truck road spacing for the system must no exceed 1 600 - 1 800 m.

Short-distance (<700 m) equipment

Winch and tower mounted on tractor, truck or trailer, working always from the road. There are two types:

a) high lead - two-drum winch, no skyline. Maximum distance 300 m. With European equipment, maximum 2,5 t load; with North American equipment, considerably bigger loads, with auxiliary drums for guylines. Interlocked lines for grapple yarding;

b) cable cranes - two or more drums, skyline, mainline, haulback line, carriage. Maximum distance 700 m. With European equipment, maximum 5 t load. A fully suspended load is possible but not practical.

Long-distance (>700 m) equipment

These systems are now seldom used as being too costly.

Possible to about 2 500 m, but feasible to a maximum of 1 500 m. A fully suspended load is required and calls for supports higher than those for short-distance equipment. Traces, supports and anchors must be calculated. The maximum load is 10 t, but in practice it is usually 5 t. There are two types:

a) all-terrain cable cranes - skyline, endless cable, carriage (with or without stopping device), winch at road with cable drive over parabolic pulley. Logs are carried vertically or horizontally. In the latter case, high supports are required.
b) gravity cable cranes - skyline, one hauling cable carriage (with or without stopping device), winch mounted on sledge working from ridge station. Logs are fully suspended, vertical and require high supports.

Long-distance transport

The mode of transport for long-distance hauling is largely determined by the existing transport facilities. Tractors can be used to distances of up to 10-15 km. The break even point between truck and railway transport in the industrialized countries is usually about 50-100 km in cases where both roads and railways already exist. The break-even point between water transport by barge or floating and truck transport is usually in the same bracket but much depends on the road standards and the suitability of the waterways for timber transport.

1.5 ROLE OF MAN-POWER IN FORESTRY

Whatever the context is for a forestry enterprise its production should not be carried out at random. It needs to be planned, supervised and controlled so as to meet some efficiency criteria or other goal. As man-power constitutes the major input it is evident that its ability to work and produce is of paramount importance for the economy of the forest enterprise, whatever its size, type and goal. It is also evident that knowledge of man as a production factor - both as an individual and as a social group - is needed. The potentials and limitations of man for carrying out the work is not only useful to know to prevent hazardous work practices, but it greatly improves the possibilities to arrange the forestry production in a manner consistent with the efficiency criteria each enterprise must adhere to, to exist and stay in business.

Man was long considered to be in a position to master his own work situation, to adapt his tools, methods of work and his work place to his own capabilities. Only the training, to acquire the necessary skills, was recognized as a need and in forestry it was long - and in many places still is - considered sufficient with some kind of primitive on-the-job-training performed from father to son or by rudimentary instructions from work comrades of foremen.

Perhaps this attitude had some relevance in days when the worker had some real possibilities to influence his own work situation, when he worked alone or in a small group in a rather simple and easily adaptable work environment. But the change from such a craft-manlike production to industrialized and highly specialized jobs has drastically changed work both in substance and form. Technology now moulds the jobs and it is now much less the worker than the engineer or forester that shape the jobs thereby curtailing the workers possibilities for the mutual adaptation between worker and job. The responsibility for such an adaptation is largely transferred to the employer and this needs to be understood and recognized.

Labour is usually regarded as one of the inputs, one of the production factors, in forestry and this approach will be maintained throughout this book. Operational efficiency is to economize with this input and to balance it against the outputs. It follows that labour inputs shall be limited in one way or the other. However, because of the spatial character of forestry, its spread over large areas with often few other means of work or production, forestry will often have strong impact on employment in rural areas, on the social conditions for living there and on the whole distribution of the population in a country or a region. Forestry then sometimes - in addition to producing goods and services for man - may be viewed as a means to promote desirable social developments and to employ people. This certainly holds true also for most other branches in national economy but forestry as being one of the few existing possibilities for work in certain areas, will have a more decisive key position in this respect. The provision of employment, one of the critical problems facing the world, particularly in many developing countries, then becomes an objective of forestry and is obviously contradictory to economizing with human inputs. But this contradiction is only seeming as always some sort of efficiency critieria should rule. One way of solving efficiency problems when social objectives are strong - such as in the planning of rural developments and creation of employment - is the <u>shadow-pricing</u> which is briefly described on page 38. Through shadow-pricing economic and social objectives may be brought in to some form of harmony.

1.5.1 <u>Man-power Planning, Employment and Labour Productivity.</u> The man in forestry can be classified in some broad categories depending on his tasks and function. Similarily, the jobs can be described with regard to the requirements on the individuals manning them. These descriptions have the scope that the profile of the workplace shall conform with the capabilities of the worker. Man-power planning is necessary not only for the enterprises but also for authorities concerned with education and training. It is one of their most needed responsibilities to supply the labour market with man-power in quantity and quality corresponding to its needs. The following educational or training levels are usually recognized which correspond to man-power requirements in most organizations:

<u>Professional</u>: generally university graduates.

<u>Technical</u>: men who have undergone formal full-time training usually for a period of one or two years.

<u>Vocational</u>: skilled men, who have undergone formal vocational training or have been promoted from the labour force and given ad hoc short-course training. Foremen, forest guards etc.

<u>Labour</u>: <u>Skilled workers</u> who have either undergone worker training or aquired skill on the job.
<u>Unskilled workers</u> who have no training for the vocation or little experience.

For the purpose of educational planning the ratios at which these categories are found in forestry are of interest. <u>Richardson</u> (1967) suggests the following composition:

Table 1.10 Labour supervision requirements in field operations (Richardson, 1967)

Type of forestry	Ratios		
	Vocational/ labour	Technical/ vocational	Professional/ technical
Non-revenue producing management	1:10	1:8	1:6
Revenue producing management	1:8	1:7	1:6
Resource creation	1:5	1:6	1:8
Timber supply	1:4	1:5	1:9

Translated to more concrete rule of thumbs one foreman can supervise 4-10 workers, one technician 5-8 foremen and one professional 6-9 technicians. This ratios should be regarded more as an guide for educational planning than as a suggestion to organize a certain activity. It also applies to the traditional way of setting up a supervisory hierarchy. But new ideas in this respect are now emerging as will be made clear later in this text.

Work in forestry is often seasonal. Some work operations can or must for biological reasons only be carried out during a certain time period of the year, such as planting and weeding. Some harvesting operations are strongly influenced by climatic conditions, e.g. deep snow or heavy rains, making such operations so costly or difficult that it is preferable to close them down. Forests are often spatially mixed with agricultural land. As the demand for labour in agriculture often is seasonal, an employment pattern often developes where a considerable part of the rural labour force, including animal and machines, works periodically in agriculture and forestry. Sometimes the demand for labour in agriculture and forestry is complementary in the sense that it appears or is concentrated to different time periods. This was earlier the case in the temperate softwood forests in the northern hemisphere, USSR,

Scandinavia and Canada and led to a seasonal distribution of the work opportunities to great advantage for the rural population, forestry and agriculture as well. In forest regions with much agriculture this annual work rythmn still prevails, in particular where the farmers own also the forests. Increased mechanization, an extended forest road network, and the use of specialized equipment mastering the jobs under any climatic condition tend to sever the employment ties between agriculture and forestry, primarily because such specialized forestry equipment must be worked as many shifts per year as possible to keep the capital cost of the machines on a reasonable level. But labour demand from agriculture and forestry can also be conflicting. In tropical or sub-tropical regions with distrinct monsoon seasons and heavy rains the work opportunities often coincide.

In particular the planting of trees takes place during periods of high demand for labour in agriculture. However, forestry techniques both in silviculture and logging are systematically developed to make them less dependent on climatic factors in order to spread the work over longer periods but often seasonal extremes in climate put absolute limits to such efforts. Thus the labour force in forestry shows up a complex composition with regard to origin, duration of work and dependency-relation to forestry. The main categories are:

Employees:	Non-Employees:
Permanent workers	Forest farmers
Seasonal but annually recurrent workers, often from the farming sector	Contractors and entrepreneurs
Occasional or temporary workers	

The subsequent text will focus on labour problems appearing in situations of employer/employee relationships but a substantial part of the content will be applicable also to other employment forms. Firms often show their quantitative input of labour by the number of workers. Unless also the number of workdays or workhours per year is given, such a figure is rather meaningless in forestry because of its seasonal character. Apart from being useful for trend studies over the years within the same enterprise, figures on labour inputs are of interest for comparisons between forest enterprises, regions and countries although cautions should be taken not to draw too far-reaching conclusions of them, unless based on a careful analysis of differences in their work situations. There are innumerable factors causing variance in labour requirements for which allowances must be made in cross-comparisons whereas comparisons over time within the same enterprise or locality are on safer grounds.

To make labour inputs more understandable, the input should be related to some forest parameters. Two such parameters are commonly used:

(i) labour input per unit area of (productive) forest
(ii) labour input per unit of production from the forest

Both these indices are useful and complementary. The first one, input per unit area, reflects employment potential of forest land and can also be informative on intensity of forestry. The second one is the usual expression on the gross labour productivity in forestry. Both the indices can be broken down on the various work operations in forestry such as silviculture, logging and transport whereby the possibilities for comparisons increase.

The inverse of labour productivity, which can be called labour consumption or labour requirements, is often used. It has the advantage that figures for single work operations can be added up to an aggregate figure.

Examples on such statistics on labour ratios are already shown in Chapter 1.1.

1.6 ERGONOMICS

Ergonomics is the science for the study of the mutual relationship between man and work. It can be subdivided in the following main components:

Work physiology
Work medicine and hygien
Functional anatomy
Psychology
Sociology
Aquisition of skills - Training
Accident prevention

Forest work for more than 100 years has been studied with regard to ergonomic aspects. There are several reasons for that. It was and often still is physically very heavy. Many individuals perform exactly the same type of work, such as working with an axe or a saw. The results have often proved to be of great practical value for finding solutions to work problems hampering efficiency or else being unwanted.

A complete coverage of ergonomics is outside the scope of this book. However, a few of the analytic methods will be briefly described which are of particular value for improving work productivity and other aspects of the work. The reader wanting to learn more is referred to Åstrand and Rodahl (1970).

1.6.1 Work Physiology. Man's physical capacity often limits his output in forest work, and consequently a reduction of the work load is highly desirable. In physiological work studies various indices are used to measure the work load. Such measurements can be carried out over time period of various length and each analytic method or index is often relevant to answer or elucidate a specific work problem.

The daily energetic cost, the energy generated in the body, as well as the specific energetic demands in different types of jobs is generally measured in calories (1 kcal = 1 000 cal = 4,19 kJ = 1,16 . 10^{-3} kWh).

The energetic cost varies with occupational activities. The heaviness of forest work is caused by the need to move around in the terrain, the carrying of tools and machines and the manual handling of heavy logs.

Table 1.11 Daily energetic cost during rest and in some occupations

Activity	Energy output (kcal/day)
Resting in bed	1 500 - 2 000
Office work	2 000 - 3 000
Industrial work	3 000 - 4 000
Building and agricultural work	4 000 - 5 000
Forest work	5 000 - 6 000

There is a considerable variation in work demands in the forest jobs and the energetic cost will also vary with the speed at which the job is performed. Some examples are given in the following table as typical averages.

Table 1.12 Energetic cost of some forest jobs

Job	kcal/min
Digging drainage ditches by hand	8.0
Planting by hand	6.5
Sharpening saw	3.2
Carrying power saw	6.5
Sawing by hand	7.2
Felling	8.6
Barking	8.0
Carrying and dragging logs	12.0

The energetic demands in many forest jobs are often so high that they limit the production rate. It is therefore of great value in the rationalization of the work and in efforts to improve the lot of the workers to know exactly the heaviness of each work element. Ergonomic studies play an important role in forest work science and some fundamental knowledge of such studies will be very useful for foresters. The energetic processes are those which permit the release of energy in the form of movement and heat. They are of two types:. aerobic and anaerobic. The aerobic process can be illustrated with the following formula:

$$C_6H_{12}O_6 + 6 \cdot O_2 \longrightarrow 6 \cdot H_2O + 6 \cdot CO_2 + \text{free energy}$$

(sugar + oxygen ——— water + carbon dioxide + movement and heat)

This briefly is what happens in the muscular tissues during the combustion of the food stuff (prepared in the digestion organs and transported in the blood) and with the oxygen of the inhaled air, diffused in the lung tissues to the blood and transported in the blood to the muscles.

When very heavy activity is performed, the oxygen supply to the muscle tissues is not sufficient. In this case, the body gets part of the required energy by anaerobic processes:

$C_6H_{12}O_6$ ——— $2C_3H_6O_3$ + free energy

(sugar ——— lactic acid + movement and heat)

However, the lactic acid has a toxic effect upon the muscle tissues, the presence of which is unfavourable for muscular work. Therefore, the anaerobic processes, in addition to the energy release, cause local muscular pain. The worker after a short period performing anaerobic work, unconsciously diminishes the activity by taking a pause. During the rest the lactic acid is oxidized to other substances more adequate for muscular work.

1.6.2 <u>Use of Ergonomic Research Methods in Forestry.</u> Some of the ergonomic research methods of particular value in the analysis of practical problems in forests work will be briefly described.

Nutrition

The intake of food must balance the high energy demand of forest workers. It can be determined by measuring the consumption of various foodstuffs during a period of time, e.g. two weeks. Based on the calorific values of the various foodstuffs, the calorific intake is computed. About 10% of the food is not absorbed in the intestines. These methods give only daily averages. If a worker does not get the necessary amount of calories, he will adapt - for biological reasons - the speed of work and the rest pauses according to the calories he has available for physical work. He has no other choice. A person in complete rest has a caloric demand of at least 1 500 kcal. In many developing countries the level of nutrition is often around 3 000 - 3 500 kcal of which perhaps 500 - 1 000 kcal must be used for home work and for the transport to the work place. Very little remains for actual work in the forests, often requiring 6-8 kcal/minute or even more. Sometimes the diet will also be a problem.

1.6.3 <u>Measuring the Energy Metabolism.</u> In work study, the efficiency is generally expressed as output per unit time. If at the same time also the energetic cost is measured, as with engines, and efficiency also computed as output per unit energy very useful additional information is aquired for job improvements.

The energy metabolism during work can be determined by measuring the oxygen uptake, one liter of oxygen corresponding to approximately 5 kcal. The inspired air is measured and the composition of the expired air is analyzed. The oxygen uptake is the difference between the volume of pure oxygen in the inspired and expired air. Special instruments are developed for such studies which can be used in the forest work.

When the energetic requirements increase the heart rate also increases. This correlation can be used for an indirect determination of the energetic cost of a work. This is a very simple and practical method of great value in forest work study. The pulse can be counted during short stops usually the time for 10 heart beats, or instruments can be used which transmits the heart rate by radio to some recording, and sometimes also computing, device. Even rather complex jobs can be analyzed in detail as the heart rate adjusts to the work load rather quickly once the worker is "warmed up".

The body temperature rises during work to a level which is dependent upon the energy expenditure and if measured during the course of the work day it indicates the average work load.

The grading of the heaviness of work can be made according to a scale worked out by Christensen (1955).

Table 1.13 The grading of the heaviness of work (Christensen, 1955).

Heaviness of work	Oxygen consumption 1 O_2/min	Pulse rate beats/min	Body temperature °C
Very low	0.5	75	
Low	0.5 - 1.0	75 - 100	37.5
Moderate	1.0 - 1.5	100 - 125	37.5 - 38.0
High	1.5 - 2.0	125 - 150	38.0 - 38.5
Very high	2.0 - 2.5	150 - 175	38.5 - 39.0
Extremely high	2.5	175	39.0

Grading of the heaviness of work is made by direct measurement (oxygen consumption) or by indirect estimates (pulse rate, rectal temperature).

It should be noted that there is in man an individual relationship between the oxygen consumption and the other two indexes depending mainly upon the physical fitness of the person. One liter of oxygen consumption corresponds to 4.85 kcal. Thus a moderate work level requires 4.85 - 7.28 kcal per minute which is equivalent to 0.34 - 0.51 kW. However, only up to 25% of this can be used for external work if large muscle groups are used. The balance is dissipated as heat. In practical work considerably less than 25% is used.

The sweating rate integrates the physical workload and exposure to heat. The weight of the person is recorded at the beginning and end of the test period, usually a shift, as well as food and water intake, urine and faeces, the balance being evaporation of water by sweating. Heat stress is a big problem in a hot and humid work environment. See Appendix A.3.

Undesirable static work can be mapped with electromyography and thus be identified and rectified.

Accident prevention is generally based on information obtained by analysing accident statistics. However, if a rather long time elapses until sufficient number of observations are obtained, this tends to make the findings obsolete. In many industries, also in forestry, the "near-accident" (or critical incident) method has been used with considerable success. A "near-accident" is defined as an unexpected event which could have led to an accident. The workers themselves are asked to record such events and relevant situation factors during some suitable work period, say 1-2 weeks. The number of observations will increase to perhaps 50 or 100 times as compared to real accidents and the findings can rapidly result in remedial actions. "Near accidents" can also be discussed between employers and employees in a more relaxed atmosphere.

There are many other ergonomic research methods such as questionnaires for problems related to work psychology and sociology. The ones described above are those which can be used by forest enterprises themselves, provided they are supervised by ergonomic or medical expertise.

1.6.4 <u>Ergonomic Studies and Work Efficiency.</u> Ergonomic studies are made to fit the job to the worker, to improve safety and health and to combate noise, vibrations and toxic agents. However, they are also very useful as a means for a more direct improvement of work productivity, especially in heavy physical work and for the design of work places, operator seats etc. In addition to the time consumption obtained in regular work studies, ergonomic studies also provide figures on other work loads such as physical or psychological strain, static work load and heat stress.

From the point of view of research design, the labour force should be looked upon as a population with a considerable individual variation with regard to skill and work capabilities. To ensure the validity of a work study a number of workers should be studied, preferably randomly selected. This requirement is often difficult or impossible to meet for cost reasons. If time studies are complemented by ergonomic research the inference of the individual variation among the worker is to a considerable extent brought under control. Experiments and studies for the choice of methods and equipment can be reduced in duration and scale as the ergonomic studies will provide additional insights on which the choice can be made on safer grounds.

Apart from carrying out ergonomic measurements in ordinary work studies, special experimenets can be made in the form of standardized tests in which only one variable is systematically varied and other variables in a forest work situation are kept constant. Such tests can then be of very short duration, less than ten minutes, for each test and the needed knowledge be obtained in short time at low costs.

An example on the design of such tests is described below.

Scope of the study:	To explore the influence of the weight of a power saw in carrying the saw in forest terrain
Design:	Select a typical path in the forest terrain and three sizes of power saws (light, normal, heavy). Carry out one test for each size of the power saw at three different speeds of walk - slow, normal and fast.
Execution:	Select a worker with experience of the work, describe the task for him but let him choose the way of carrying the saw. Instruct the worker to maintain the same tempo throughout each test - low, normal and fast. Begin the test with a warming up period of around four minutes after which the measurement of speed of walk and physical work load is made during 2-4 minutes.
Measurements:	Length of walk, time Oxygen consumption or pulse rate
Compute:	Spead of walk, meter/minute and oxygen consumption (l/min) or pulse rate (beats/min) and plot the figures in graph, A. Draw the curves. Read in the graph B: Output at equal work load (full line) Work loads at equal output (dotted line)

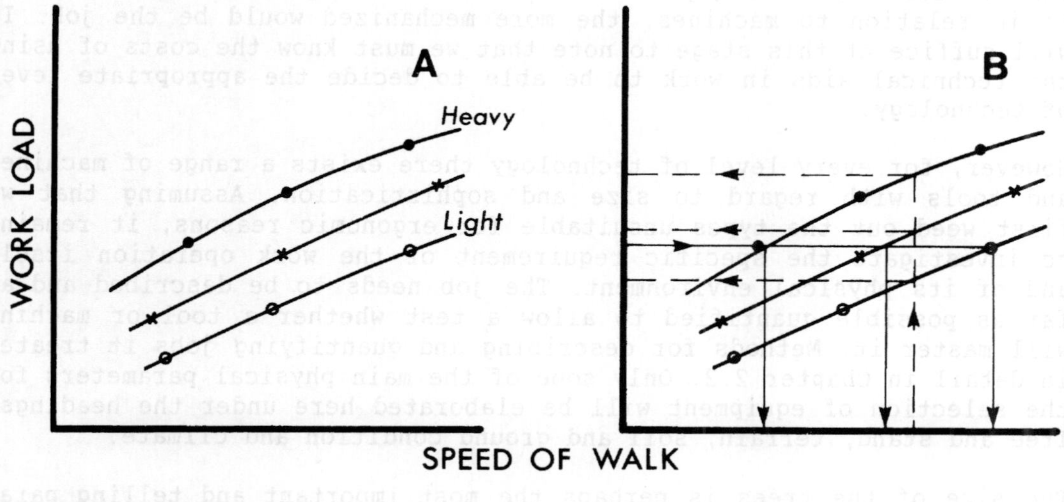

Figure 1.7. Graphs illustrating the influence of the tempo of work on the physical work load of the worker carrying a power saw (graph A) and how outputs at equal work load or work load at equal output can be assessed (graph B).

If one wants to know the influence of the slope, snow depth or any other variable the same study is repeated for each wanted parameter value. Comparisons can then be made between all tests at the same physical work load or at the same speed of walk.

Standardized tests of this type can be made for practically any method or equipment. It does not fully correspond to a true work situation in the forest but will provide useful insight into the work process, through which practical experiences can be better understood and improvements be made.

Not only manual work can be studied in standardized tests with different speeds of work. Testing machines in the same way can often yield results contributing to a better understanding of their performance under varying outer conditions.

1.7 MAN-MACHINE SYSTEMS AND WORK ENVIRONMENT

It is now common to talk about a man-machine or man-tool system to denote the interdependence between the worker and his technical means. It is actually the efficiency of such a system that decides the rate of production in any job. It is obvious that just as we need knowledge about the worker's capacity in work, we also must know exactly how suitable and useful are the tools, the equipment or machines which are available or can be put at his disposal. One such aspect has already been dealt with in Chapter 1.6 Ergonomics, namely the importance of selecting technical aids fitted to the worker's physiological capacity.

But the selection of technical aids must also involve adaptations to the work environment and to the work object. The first choice is concerned with the level of technology, the decision whether or not machines shall be used and how sophisticated these machines shall be. Machines are almost always used in order to substitute for manpower and it is therefore obvious that the relative cost of man-power and machines will bear strongly on the choice. The more expensive man-power is in relation to machines, the more mechanized would be the job. It will suffice at this stage to note that we must know the costs of using the technical aids in work to be able to decide the appropriate level of technology.

However, for every level of technology there exists a range of machines and tools with regard to size and sophistication. Assuming that we first weed out the types unsuitable for ergonomic reasons, it remains to investigate the specific requirement of the work operation itself and of its physical environment. The job needs to be described and as far as possible quantified to allow a test whether a tool or machine will master it. Methods for describing and quantifying jobs it treated in detail in Chapter 2.2. Only some of the main physical parameters for the selection of equipment will be elaborated here under the headings: Tree and stand, terrain, soil and ground condition and climate.

The size of the trees is perhaps the most important and telling parameter for a logging operation. It is usually expressed in diameter breast height (DBH) or (merchantable) volume. From the tree size also log size can be assessed once log length is decided upon. But it is important to know not only the mean or average tree/log size but also its range. In particular the size and frequency of the big trees is important as these create the biggest technical problems and also represent the greatest values.

It is so that certain technologies are developed for a certain (average) tree of log size with equipment designed for it. When such equipment is put to work on smaller or bigger trees its efficiency generally goes down. There is also a general rule that the smaller the trees the lower is the production rate.

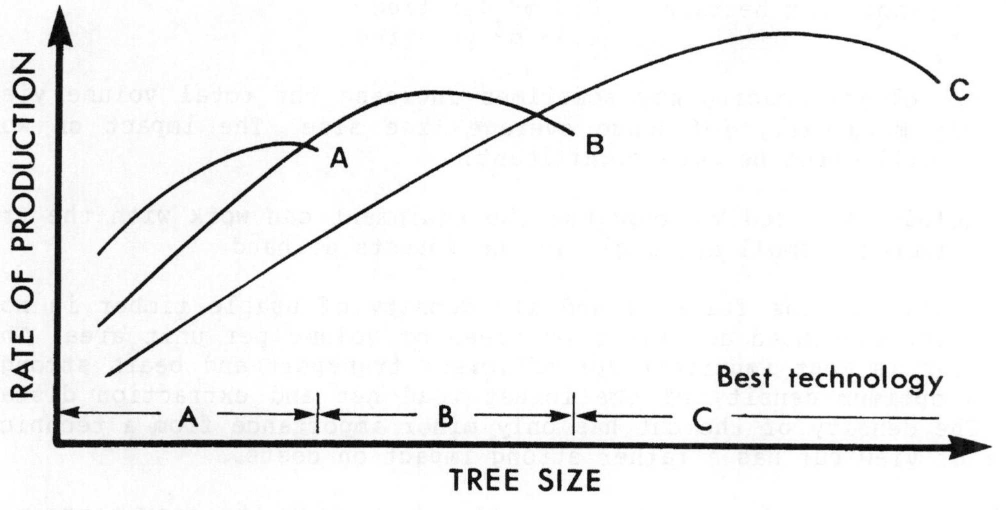

Figure 1.8. Typical correlation between tree size and rate of production for three alternative logging systems, A, B and C.

The relations between tree size and production rate apply to both tools and machines. The energy for sawing with a manual saw is consumed by the friction between the saw and the wood and by cutting off the fibres and transport of saw dust out of the kerf. The friction is closely related to the number of saw teeth at the same time working in the wood. If this number is high - if the saw teeth are closely spaced - much of the energy a sawyer can exert is lost in friction and little for cutting fibres and transporting sawdust. Everyone who has tried to cut off a big piece of lumber with a carpenter's saw with closely spaced teeth, designed for small sizes of wood can verify this. And the same general rule of adaption to tree size applies to axes, power saws, tractors, loaders, high-lead yarders, chippers and practically any kind of equipment.

The decline in production rate with decreasing tree size is the reason and explanation why trees below a certain size cannot economically be harvested. The low production rate leads to a high cost per unit volume, exceeding the value. However, it should be observed the minimum size tree may very well be quite different for different harvesting systems.

It should always be kept in mind that the average tree size during a rotation of a stand is determined by the initial spacing at the establishment or after a precommercial thinning. Thus

$$\text{Average tree size} = \frac{\text{total production per unit area}}{\text{number of established trees}}$$

If a specific site yields 600 m^3 during a rotation the average tree size will be

at 2000 plants per hectare   0,3 m^3 per tree
at 4000 "-   0,15 m^3 per tree

However, closer spacing may sometimes increase the total volume yield but only marginally influence average tree size. The impact on wood quality will often be more significant.

To conclude, we need to know how the equipment can work with the tree size - average, small and big - in the forests at hand.

With regard to the forest stand its density of usable timber is most important, expressed as number of trees or volume per unit area. This parameter is most important for efficient transport and bears strongly on the optimum density of the forest road net and extraction distances. The density of the cut has only minor importance from a technical point of view but has a rather strong impact on costs.

As forestry in most countries is pushed back from the easy terrain to less fertile and more difficult topography it is indispensible to know how equipment can master difficult terrain conditions, such as steepness and pattern of topography. There are two critical values on steepness. One is the angle of repose of the soil, which usually is around 70%. In terrain steeper than 70% road construction becomes most difficult, expensive and harmful for the environment as fills cannot be made. It simply slips downhill and the road has to be built on a shelf cut into the slope. The excavated material should be endhauled instead of side casted. The other critical parameter on steepness is the maximum slope below which tractors can work satisfactorily. It varies between 40-60% depending on type of soil and its moisture content but 50% can be regarded as a good average for the forest tractors in use in the mid 80's. The steeper slopes must be harvested with cable systems or sliding down the logs by gravity on the ground. Both these practises are considerably more expensive than tractor logging and - apart from technical difficulties to thin the stands with these techniques - thinning operations are often economically unfeasible.

There are two influences of the climate of immediate importance for work efficiency, the influence of percipitation on soil or ground conditions and of temperature, air humidity and insolation on the comfort of the worker. In cold climates snow depths over 0.5 meter create increasing difficulties, especially for workers on the ground who have to move around in the forests. Modern forestry machines can with some difficulty master snow covers of 1 meter or more. Heavy snow fall or snow drift may create problems to keep the forest roads open. But moderate snow covers and frost are often advantagous for transport operations. In temperate or hot climate rainfalls - either occuring as very heavy short rain storms or long periods with continous precipitation create a number of technical problems such as reduced bearing capacity of the soil, road beds or road surfaces due to high moisture content, erosion and sometimes also waterlogged forests. Extreme dry weather periods may create such fire hazards as to require the prohibition of smoking and sometimes even the complete closing down of work activities in the forest. Dust abatement may also be required on the dirt or gravelled forest roads to prevent the deteoriation of the road surface through loss of the fine soil particles necessary to maintain a smooth road.

Work in the open in rain is often rather unpleasant and if it lasts more than some hours forest workers are inclined to call it a day. However, there seems to be a considerable adaptability as people who live in rainy regions care less about getting wet and endure longer.

However, the most important comfort factor is the influence of the climate on man's heat regulation of his body temperature. Very cold climate does not pose any serious problems, except at very low temperatures, as protection against cold can be provided for by adequate clothing. But high temperatures, especially if combined with high air humidity or insolation may seriously impair man's ability to work. Heat problem can arise already at air temperatures of 25°C. This particular problem is dealt with in more detail in the Appendix A.3.

High altitudes exceeding around 2 000 meters also impair man's capacity for physical work and this is true also for people who live under such conditions.

Just as the work situations in forestry show a very wide variance in the sense that factors of considerable technological importance combine in an infinite number of ways, there are also a very large number of possible techniques to match these complexities and situations. The adequate technique can only be found if the work situations are identified and if the special technical, economic and social requirements of the situation correspond to the inherent capabilities of that technique. Such a choice was in the past mostly made on a trial and error base - man-machine systems were simply put on practical tests often in a large scale. This is both a primitive, costly and time consuming method. The task of the forester or engineer of today is to make that choice intelligently by systematic use of existing knowledge. And this knowledge no longer derives only from practical experience but also from systematic research, analysis and quantifications of situation and system characteristics. This chapter is intended to describe briefly what we need to know to solve the problems of work efficiency and to outline possibilities how to obtain that information.

1.8 ECONOMIC PARAMETERS

The efficiency of any activity presupposes that a criterion exists. Initially it is computed in physical units. There are many such criteria, each serving some special purposes. Labour productivity is one already mentioned. The disadvantage of such a battery of criteria, many of which may have validity or significance in one and the same situation, is that they are not comparable, they are quantified differently and can therefore not be aggregated.

The standard procedure to overcome this difficulty is to use money as a common denominator. If monetary values are accorded to each input or output quantity, the efficiency can be expressed in one simple number, in one single unit of measurement. The art of monetary valuation is one of the key points in Economics for which there are numerous excellent textbooks. The reader should be familiar with basic principles of economics. Only a few points of relevance for work efficiency problems will be discussed here.

The first one is that often not all inputs or outputs can be quantified - neither physically nor in monetary terms. In any estimate some known used-up resources always remain, some outputs or some consequences to which realistic values cannot be allocated, so called imponderables. As economic criteria therefore does not hold full and complete authority in making the decision, common sense must be used to assess the significance of such imponderables.

A second point is that there are different basis or points of departures for placing values on resources or assets so that the same asset or resource may be allowed different absolute or relative values depending on the base of valuation. One common case is society's valuation (or social cost) as compared to valuation of a private enterprise (or market price or cost) of labour or equipment which is further dealt with in Shadow Pricing, Appendix C to this chapter. Efficiency in monetary terms therefore depends on the context in which it is studied.

However, there can be no doubt that the most universal way is and will be to use money as the common denominator in assessing efficiency. We need economic parameters on the values of resources involved in forest work operations. To collect such parameters may not cause serious problems - but often proves rather laborious - once basic knowledge of economics is at hand.

Only formal mention will be made that there are everywhere laws, regulations, norms and time-honoured customs on what can or cannot be done and this applies in particular with regard to employment of labour and the use of the forest resource. It goes without saying that any violation of such institutions must be avoided. What we need to know is exactly the degree of freedom each society has set up. Those available in writing should pose no problem to collect. But customary rights, traditions, taboos and even predelictions may set limits on what is feasible in organization of forestry and they are often as difficult to map as to change.

1.9 METHODS OF ESTIMATING PRODUCTION with special reference to developing countries.

1.9.1 General. Logging production is affected by many factors some of which are given in the checklist in Appendix B to Chapter 2. Suggestions are also made in that checklist for measurement or description of such factors in terms relevant to logging problems. A survey along the lines suggested in the checklist would serve as a basis for the initial selection of the logging systems, working methods and equipment to be further investigated.

The silvicultural regime on one hand and harvesting systems and equipment on the other are mutually dependent and should be analysed by this dependency in mind. However, as silvicultural problems are outside the scope of this manual, such an overall approach cannot be fully maintained here. The most important aspect of silviculture for loggers is the need for thinnings and the harvesting of small-sized wood arising from the regime. These operations are always more difficult than the final fellings and the harvesting cost may often exceed the revenues.

It is imperative that silviculture and logging are coordinated by the respective expertise in order to conform with the overriding objective.

Two problems are of fundamental importance. Firstly, the spacings in planting should not be so close that small-sized wood that cannot be harvested at economic prices is produced in large quantities. Secondly, as the thinning of stands in very steep terrain is now very expensive, care should be taken in applying a silvicultural regime requiring commercial thinnings. Experience has shown that such thinnings are often neglected as being too costly. Consequently the planned regime is thrown out of balance.

1.9.2 Production Ratios and Levels. Estimating the production in logging and transport is difficult as a large number of factors affect the output in various work operations. These conditioning factors can be grouped in three categories - the work object, the work environment and the "human variables" related to the labour force.

The available studies of logging operations indicate that the influence of the first two categories, the physical variables, is rather uniform in the sense that for a certain work method and certain equipment a change in one the factors, say tree dimension, has the same relative impact on production for all types of labour. Human variables on the other hand are much less predictable and much more complex to analyse.

Nevertheless, logging production may feasibly be estimated in the following three steps when transferring experiences from one situation to another, e.g. from a developed to a developing country:

a) Establish production curves or correlations between the physical variables and output on the basis of existing documentation and data;

b) Scrutinize the characteristics of the labour force, the human variables and adjust the level of the production as derived in step (a) to the anticipated influence of the differences in human variables between the actual labour force and the labour force from which the data derive;

c) Check, if possible, by means of a work study that the established production level in step (b) is appropriate for the project case. Such checks can quite easily be made as it suffices to select only one or two points on the production curve established in step (a), relate the values so established and compare the ratio with the adjustment according to step (b).

1.9.3 <u>Human Variables.</u> Variables related to the labour force are numerous and only a few can be included here, such as skill and training, physical work capacity, nutrition, health, heat, altitude and some psychological factors. The term human variables is not quite adequate as some are actually situation factors. They do, however, bear on the worker's ability to perform a job.

It is not likely that the aggregate influence of these factors will equal their sum as interaction among them is almost certain. As the result of such interaction escapes accurate measurement, in particular with regard to the presence of both undernutrition and heat stress, only some options are outlined below in an attempt to evaluate their aggregate influence.

The following procedure is suggested:

a) When the time consumption for a job is given in effective working time, an allowance of about 15%, unless otherwise stated, should initially be added to account for short pauses for rest during the shift time.

b) The skill of the labour force is then considered. Skill should be seen in the light of two factors - training and the labour turnover. Labour turnover may be affected by the social, economic and cultural conditions in the area but also by the occurence of seasonal employment. The following approach is suggested:

   Make the estimate for an operation where a "steady state" level of production is reached. Treat the reduced production during the running-in stage as an initial investment cost. The steady state level should be established with due regard to the labour turnover to be expected in the long term.

   In situation where logging operations are already going on in the region, these should be used as a basis for assessing the level of output and how it is likely to develop over a period of time as a consequence of increased experience and improved skills.

   Consider the need of training and the impact of the training programme suggested for the operation. The following table is suggested as a guide. However, it should be emphasized that as it is based on rather scanty observations, it should be used with care and common sense.

Table 1.14 Expected output in starting up logging and transport operations as influenced by acquisition of skill and experience (in percentage of output of experienced crew-steady state)

Training input	Period after starting up in months			
	0 - 6	6 - 12	12 - 24	24 -
	Percentage of output of experienced crew			
Small	50	50 - 60	60 - 70	70
Moderate	40 - 60	60 - 70	70 - 80	80
Intensive	60	70	80	90

Note: If the recruited labour force already has experience of other forest work, the acquisition of skill may proceed somewhat faster.

c) Consider the body size of the worker. As a body weight of about 70 kg is normal for forest workers in developed countries, make allowance for body weight by applying the factor

$$\frac{70}{\text{actual body weight in kg}}$$

to time consumption (or the inverse factor to output). This adjustment should be applied to all jobs or phases of jobs where the physical work capacity of the workers limits the work output, for example, all jobs where the workers move in the forest by foot in performing a rather heavy job. It does not apply to machine operators on tractors and trucks, who seldom leave the cabin.

d) Assess the adequacy of the level of nutrition (see Appendix A to Chapter 1). If necessary, decide on the practicability of measures to improve the diet of the workers. Check for taboos and local habits related to food. Evaluate the resulting level of nutrition and compare with the energy demand of each specific job.

e) Study the level of health in the area (see Appendix A to Chapter 1); consider measures for improvement and evaluate the impact of health on work output.

f) Collect data on temperature and humidity in the work places during relevant work seasons. Consider if the work day can be scheduled to reduce influence of heat stress and evaluate on the basis of Appendix A to Chapter 1 the influence of heat stress on work output in heavy manual work operations. Consider in this context allowance for undernutrition, item (d) above. It seems reasonable to apply the allowance for nutrition or heat stress, whichever is the higher and possibly account for the other to some extent. (For operations in hot climates the FAO publication "Heat Stress in Forest Work", Axelsson, 1974, will be an useful guide.)

g) If work places are located at altitudes higher than 1 200 m, make allowances in time consumption for physically heavy jobs. At 2 300 m work capacity is impaired by about 15 percent and at 4 000 m by some 30 percent. After adaptation, this decrease is cut by about half but this adaptation does not occur in workers who occasionally have to perform their task at high altitudes, while still living at low ones.

h) Survey the general attitude of the local population toward paid work and its previous experience of industrially organized and disciplined work. Consider allowance already made for running-in and for labour turnover; item (b) above. Decide on the practicability of an incentive or bonus system for payment of work. Assess the impact of work motivation on the level of work output as influenced by incentive systems and the cultural attitudes of the population toward paid work.

2. PROBLEM ANALYSIS

2.1 CRITERIA

This chapter will deal with the problems of efficiency. The approach will be to analyse, to measure efficiency in quantitative figures and to synthesise the findings for planning and control. To measure efficiency criteria which correctly and truly reflect the priorities and preferences are needed. As far as possible, such criteria should be quantitative rather than qualitative and for that benefits and sacrifices, outputs and inputs need to be measured or assessed.

The goods and services of the forest resource are manifold. They can broadly be grouped in productive, protective and social. Almost any forest resource includes demands of man for all these kinds of benefits. The manipulation of the resource should therefore aim at yielding a combination of such outputs, often termed multiple use managements. There are cases when the protective and social outputs are the main objectives such as park-like forests around densely populated areas and forests close to the climatic border zone of forests. But a very dominating share of the forest resource is used and will be used for the production of timber, wood fibre and fuel wood. There the protective and social outputs do not necessarily play a secondary role but as they usually can be achieved with only minor modifications or adaptations of the manipulation for productive purposes, emphasis will be on what now is often termed production forestry. In such forestry the protective and social demands are often recognized as limitations and constraints in manipulation schemes aiming at high yields of timber, fibre and fuel.

Every enterprise aiming at survival and development must by necessity strive for efficiency. The most important task of the management is to plan and control the activities so that efficiency is achieved. Many times, efficiency can be judged on the basis of previous experience or on common sense, but these are often rather blunt tools. Sharper ones are needed. The benefits and sacrifices are often composed of components of different kinds which need to be converted to the same unit of measurement so that a numerical expression of efficiency can be computed. Money is the most commonly used denominator but other ones are also possible, such as power or energy. It should be observed that the estimate of efficiency in money is not a method. Money is only a scale of measurement, a way of aggregating different units into one measure.

All problems of real life situations at first sight look complex. A large number of influences can for each problem be listed as having a bearing on the problem and many times it may appear rather difficult or even impossible to visualize a solution. However, it should already now be emphasized that it is not always necessary to tackle a problem in its full context. Many times it may prove useful to take out pieces of the problem, study them separately and put them together again. No study, whatever its scope, can claim to be complete. A fully elucidative study simply does not exist. One must be content with representations, which conform with reality as closely as is deemed desirable for each specific case. The results of a study should consequently never be blindly applied but only form a better and safer basis for actions.

Thus, the procedure in tackling an operational problem starts with the analysis. Influences, which on good grounds are considered of minor importance are ignored and those remaining are joined together in an abstraction, in a model where the different influences are given certain roles and are assumed to concur according to certain rules or even laws. In other words, a theory for the problem is formed and this theory then more or less automatically outlines the solution of the problem.

Such an approach for studying operational problems in forestry is not unique. It is used in studies in almost any context - in academic studies as well as for real life situations. Often, one may meet arguments that such theories - as they do not conform with reality - are useless. Such arguments, if relevant, are concerned with the formal deficiency of the study and have nothing to do with the usefulness of a theoretical analysis, of studies of abstractions of real life situations in the form of models.

Operational problems in forestry will in this chapter be elaborated using to a considerably extent the approach as outlined above. Do not expect to find a number of facts of immediate practical use, because your disappointment will then be great. Emphasis will be on theories and principles, which are universally applicable. Facts are many times useful and in the final stage absolutely necessary but their great disadvantage is that they are too numerous, so numerous that only a very few can be remembered and stored by the human brain. Theories and principles are fewer, they can be learned and remembered. They are, to cite Duerr (1960), "the key of the master to a wealth of facts and can easily be carried in the pocket". The difficulty with principles is not to learn them but to learn to use them for answering questions raised both in practical forestry and in forest research.

2.2 WORK STUDY

The input of man-power in forest operations is recorded through the management and control functions which each enterprise maintains. Statistics on labour productivity and similar gross data as demonstrated in the previous chapter can be computed from the accounting records of the enterprise. However, during the last half century special analytical instruments have been developed for more refined measurements of work processes for situations when more detailed information is required. Experience has shown that those instruments are not only useful for increased and more detailed knowledge of work but they have become a very powerful - and therefore also a somewhat delicate - tool for improving efficiency through new work procedures and equipment. They are also much used for establishing "a fair days work", setting of piece rates and similar managerial tasks. All these instruments can be encompassed under the term "Work Study". Numerous textbook are published on the subject matter with somewhat different emphasis and terminology. Work Simplification, Time Study, Motion Study, Work Measurementes are only a few of the most common terms.

Work Study is a systematic analysis for measuring and recording procedures, inputs and outputs in a work process. A wide range of techniques for work study exists, each developed to elucidate one or a few particular aspects of the work. These techniques will be briefly described here. The subject matter "Ergonomics" is closely related to Work Study as also some problems of mechanical engineering. Work Study can usefully be split up in two big groups - method study and time study (or work measurement). Method study has the scope of mapping or recording the various elements in a work process, generally for suggesting more simple or easy ways of performing it - work simplifications. Quantitative measurements of time is primarily not done and emphasis is laid on the sequences of work elements, movements of both workers, hands, legs, machines and also material. In time study emphasis is laid on measurement of time consumption for performing a job. It is usually preceeded by a method study or a check that the work procedures are satisfactorily organized.

Work Study can be carried at different levels of precision - the work process can be recorded or measurements made on different levels of refinements. A scale of instruments are available for these levels which can be systemized as in table 2.1. Reference is made to textbooks in work study for a description of the methods.

Systematic work study is a child of industrialization. Perhaps the most significant consequence of industrialization was that time became a more important parameter in the organization of work. Factory work was organized in shift hours. Production in series eliminating idle time became common and work simplification of work procedures to weed out unnecessary work elements was systematically applied. Even if time was a matter of consequence in the pre-industrial era, it seems that other measures were just as much used to quantify work efficiency and to organize work. For example, in the Swedish steel industry up to about 100 years ago fuels (or energy) were used to assign a day's job to the miners or steel workers and to measure their efficiency. The worker was given a certain quantity of charcoal per day and his effectivness was measured in units produced per unit charcoal. It was more or less left to the worker to decide on the time spent on the job. In agriculture workers were often assigned a task rather than to keep prescribed shift hours.

After more than 100 years of decreasing relative costs for fuels and energy, the world is now experiencing a reversed trend. Energy may again become a job characteristic of considerably stronger importance for work organization. This is one of the reason for including the chapter 4 Energy in this book.

Occasional work studies in forestry were carried out around the turn of the century and in the 20s and 30s permanent institutes were successively established in many countries. They are now an important sector in most forest research.

Table 2.1 Work study techniques, classified according to levels of precision recording and measurement

	Degree of precision: ← Macro		Micro →	
Method Study	Network Analysis Transport Flows	Process Charting Activity Charting Flow Process Charting	Product Analysis Charting Man Analysis Charting Work Activity Charting Operations Charting Multiple Man Charting Man/Machine Charting	Memo-motion Studies Micromotion Studies
				Motion and Time Measurement - MTM
Time Study	Ordinary managerial control of work days and work hours Gross Productivity Analysis	Gross Data Techniques Time Recording	Stop Watch Studies Continuous Time Recording	Chrynocyclo-graphie Cycleographic

A special time study method not included in Table 2.1 is Work Sampling. Instead of actually measuring the time consumption for each phase of work, observations are made at predetermined moments on whatever activity the man or machine is performing exactly at that moment. These observations should be randomly spaced in time but if the work cycles are irregular, fixed intervals between the observations are acceptable although it does not conform to the principles of the random sampling method. The total working time of the job is broken down in elements according to the number of observations for each such elements in relation to the total number. Work sampling can be applied on all levels of precision and for studies of widely varying durations. It is an inexpensive cheap method of observing a great number of subjects at low cost and is therefore increasingly being used.

The term "Work Sampling" is somewhat misleading. All work study is sampling in the sense that it represents the study of a job from a population of existing planned jobs. It is naturally of fundamental importance how this selection is made. To meet scientific criteria such a selection, such a sample, should be made at random but this is not always possible. In fact, the selection is often made on some practical basis with the notion that the work conditions, the crew etc should be "normal". The question, whether a study really is a reasonably true representation of reality is covered by the concept "the validity of the study" and this is a question which always should be asked in making conclusions and decision based on its results. Another concept testing the quality of a study is "reliability", exploring the accuracy of the measurements and observations. This test is recommended in applying the results of a study.

Another technique in work study is "performance rating". Man is a biological creature and not even identical twins are identical when it comes to performing a job. The problem then is whether a worker or a group of workers really represents the population of workers on which the results of the study is to be applied, in principle a "validity" issue. For practical reasons the selection of workers for a study can seldom be made randomly and for cost reasons the size of the sample often is much lower than what is statistically required for a desired level of "validity". Performance rating is developed to overcome this dilemma. The performance of each worker or sometimes a team of workers is rated in relation to some predetermined "normal" concept, e.g. normal speed of work. Such "normal speed" may in turn be based on a large number of observations of a worker population or may simply represent the work study man's own empirical experience of what is "normal". On whatever grounds the "normal" level is established, from which the performance rating is made, the crucial thing is that a subjective element is introduced on objectively measured data, outside of possibilities for control and repetition of the experiment. The study no longer meets scientific criteria and its results can be challenged scientifically. Also observation data changed or manipulated on the basis of subjective judgement no longer are statistically valid and are, strictly speaking, no longer fit for statistical analysis. Advocates for performance rating sometimes claim that as work study results are to be used for purely practical purposes, their scientific correctness is less important and it is their practical value that counts. Critics on the other hand - to which the authors of this text belong - claim that it is exactly results for practical use that do qualify for scientific, objective criteria and the more "practical" the stronger are these demands. Work study has stirred a lot controversy in many workplaces and innumerable are the conflicts and strikes causes by them. In not a few cases performance rating is at the heart of the matter.

One approach to deal with the problems of performance rating has been developed in Scandinavia by the design of "comparative studies", the principle being that a worker or a team of workers are not representing a population of workers but they are compared with themselves. The biological variation in man is thereby excluded. The methodology is best explained in an example. Assume that it is required to establish piece rates for pulpwood felling under a variety of stand conditions. The approach is then that forest firms in the regions for which these rates shall be applied are asked to compute from their records the daily production per worker. Assume that the average production per worker turns out to be 10 m^3 and the average earning 40 US$ corresponding to an average piece rate of 4 US$ per m^3. These data are then supplemented with a number of tree and stand parameters as well as data on the work environment (say terrain, underbrush etc.). Assume that the average tree size is DBH 20 centimetre. Then, a set of questions are put to the work study, e.g. what is the influence on production of tree size, steepness of the terrain, branchiness, underbrush etc. The work study team then designs an experiment in which each worker carries out the felling on samples plots differing only with regard to tree size. Say that for each worker sample plots are selected with average DBH's of 10, 15, 20 and 25 centimetre. The production of each worker on these four plots is the measured in time studies and ratios of production as related to DBH for each worker established individually. Say that three workers are studied and the production rates are as follows, the ratio for the 20 centimetre stand being 1,00.

Table 2.2 Example from a work study of production rates as related to tree size. For each worker the rate is set to 1,00 for a tree of DBH = 20 cm.

	DBH, centimetres			
	10	15	20	25
Worker 1	0,55	0,85	1,00	1,12
2	0,51	0,81	1,00	1,16
3	0,54	0,82	1,00	1,15
Average	0,53	0,83	1,00	1,14

As these ratios are rather consistent the following conclusions can be made (if the results are not consistent further studies are needed):

If the average earnings of the labour in this particular region is to be 40 US$ per day the piece rates for pulpwood felling as related to average DBH of the stands should be

for DBH 10 centimetre 4:0,53 = 7,28 US\$/m^3

15 " 4:0,83 = 4,82 "-

20 " 4:1,00 = 4,00 "-

25 " 4:1,14 = 3,50 "-

It is then left to the negotiating parties - the forest firms and the workers of the labour union - to come to an agreement with regard to "a fair days pay", whether it should be 40 US$ or something else. The influence of each tree, stand or work environment factor can then in principle be explored through studies on samples plots which differ only in one factorial aspect. However, the same factorial analysis can often be made by the use of regression analysis or comparisons between single work elements, e.g. debranching of trees with varying crown length, whereby the design and extent of the study can be greatly reduced. No subjective judgements are made and both the validity and the realibility of the results can be tested with standard statistical methods. The clue is that work study refrains from establishing levels of production and limits its task to ratios of production under various sets of condition.

2.3 THE FOREST RESOURCE

2.3.1 Tree, Stand, Location and Accessibility. Goods and services from production forestry are in demand by people who generally live remote from the forest resource. Even when they live in or close to the forest they require timber and wood fibre in a processed form requiring transport of the logs to mills where such processing is made. The distribution of the processed goods to the places of consumption is also required.

The location of the forest resource is consequently decisive for its usefulness. Similarly, the location of the processing industry - for which there is a free choice at its establishment - is of utmost importance. Our first concern will thus be to explore the impact of location or more specificly the implications of distance between the resource and the processing industry and ways and means to master this distance by means of efficient transport.

For various reasons, which shall be further examined later in this text, the logs or wood fibre are carried from their place of origin to the market by various modes of transport forming a chain of successive transport operations. Each link in such a chain consists of a subsystem of equipment and man-power for the actual movement of the goods.

The subsystem also includes transport routes on which this movement is carried out. Furthermore, there must be places where the goods can be transferred from one mode of transport to the other, requiring space, manpower and generally also equipment. The places for such transfer are called terminals. All these assets form a transport system. Its objective is to provide as cheap access to the resource as possible, permitting the movement of goods at a low cost.

The use of the forest requires that labour, machines and often also tree plants, fertilizers etc are taken in, wood brought out. The routes for the transport of timber are therefore often used for a movement in a two-way traffic and the cost of these transports can be considerable. Consequently, in the design of the network of transport routes both timber, labour, equipment and other material must be duely considered

Methods of assessing accessibility of the forest resource is further elaborated in Appendix

The infinite variation in work conditions and work content in forestry can perhaps be best described by the recently coined term "inhomogenous homogeneity". It connotes the large number of physical factors, their great variance and the resulting infinite situations by a largely randomized combination of them.

In contrast, the character of a factory, say a steel mill, a pulp mill or any machine shop is standardized homogeneity. Work content, work pieces, products, equipment etc is purposely standardized in an utterly uniform and controlled environment. The nearest forestry can come to that are roads built to prescribed standards and monocultures managed without thinnings to produce trees of equal size and shape for the final felling. But nature itself severely upsets any such efforts.

Inhomogeneity has to be accepted in forestry. But the homogeneity of it implies a certain degree of order and conformity that can be used to get at least some degree of control and streamlining which is so important for efficiency in any production.

This has to be understood when attempting to measure, classify and describe work factors in forestry. First, one has to rely on sampling and the whole set of tools in statistics, such as averages, mean values, standard deviation, quartiles etc. Also extreme values, mostly on the higher side are important. These are entities on which production records and norms are based and which must be used in the forward planning.

2.4 INTRODUCTION TO TRANSPORT THEORY IN FORESTRY

2.4.1 <u>The Components of Transport</u>. In order to explore some of the most significant features of a forest transport system, its components must be systemized along similar lines as in transport economics in general. It is convenient to distinguish items or components related to

(1) Planning, supervision and administration
(2) Movement of timber, labour etc.
(3) Terminals
(4) Transport routes

In the end, the cost of the whole transport system must be born by the merchantable products or services for which it is designed. Some work operations or cost components are directly related to the actual movement of the products. An example is that the cost of driving a truck loaded with timber from one point to another can be directly charged to the actual load. One can thus talk about direct transport work or cost, as it can be directly associated with a known volume of goods. On the other hand, the road on which the truck travels is used by other timber trucks and other vehicles moving labour or other items. The cost of this road cannot be directly associated with the movement of any specific load but only indirectly charged to each road user through some specific rules or procedures.

A systematic division of a forest transport system, based on its functional criteria, will therefore be:

Work component	Cost component	Abbreviation
I Planning, supervision, control, administration	Overhead cost of transport	OC
II Movement of goods		
a) over a distance (transport proper)	Variable direct cost	VDC
b) at terminals (loading, unloading)	Terminal direct cost	TDC
III Installations		
a) transport routes	Variable indirect cost	VIC
b) terminals	Terminal indirect cost	TIC

Note: There is no generally accepted transport terminology. The term "variable" implies in this case that it varies with or is dependent on the distance. The terms "distance cost" or "underway cost" are sometimes used. It would then be logical to use the terms "fixed" or "constant" for the terminal costs but this can easily be somewhat misleading. "Direct" and "Indirect" imply that the costs can - or cannot without a special assessment - be charged to the transport goods.

Using these symbols the total cost of a forest transport system, TC, can be written as follows:

$$TC = OC + VDC + TDC + VIC + TIC \tag{2.1}$$

The objective of a transport system is to carry out its function in an efficient manner, - or using a cost criteria, at the lowest possible cost. Often the term "optimizing" of the system is used in the sense that a combination of the system components is sought giving the lowest possible cost per unit product. The approach in the following will be to explore and analyse these components, followed by a synthesis by which optimum systems can be designed. The "Overhead Cost of Transport" offers little interest from an analytic point of view and will be dealt with rather briefly.

2.4.2 Variable Direct Cost of Transport. To move goods over a distance requires the input of manpower and usually also equipment, e.g. a truck with a driver. The capacity of such a transport unit is determined by its speed and its load. Assume that the truck can carry 20 tonne and travel at 40 km per hour, its capacity is obviously

$20 \cdot 40 = 800\ t \cdot km/h$

However, it is peculiar for forest transport that the transport units must travel empty on the return. Thus, the capacity of a roundtrip is half (400 t · km/h) with the same speed unloaded as loaded - more if it travels faster empty and less if it travels slower.

In Table 2.3, the transport capacity of some common transport modes in forestry is shown, together with some other modes of general interest. A close look at the figure reveals the very great variance in capacity for the different modes of transport. (See the column for Ratio, where the capacities are expressed as multiples of "Man, dragging logs".)

Table 2.3 Examples of capacities of modes of wood transports for roundtrip (returning empty).

Means of transport	Payload tonne	Speed km/h loaded	Speed km/h empty	Capacity, t · km/h roundtrip	Ratio
Man, ideal				0,12	× 3.3
dragging of logs	0.06	15 m/min	30 m/min	0,036	× 1.0
Horse					
skidding	0.7	30 m/min	40 m/min	0,72	× 20
hauling	4	80 m/min	70 m/min	8,96	× 250
Tractor					
skidding	4	80 m/min	120 m/min	11,52	× 320
forwarding	14	40 m/min	60 m/min	20,16	× 560
Truckhaul	32	60 km/h	70 km/h	1 033	× 28 700
Watertransport					
rafts of truck bundles	10 000	1.6 knots	12 knots	26 146	× 726 300
ocean rafts	16 000	1.6 knots	12 knots	41 834	× 1 162 000
barges: small	2 000	6 knots	10 knots	13 890	× 385 800
large	15 000	14 knots	16 knots	207 378	× 5 760 500
Helicopters					
large	4-8	160 km/h	160 km/h	320-640	× 8 900-17 700
Cyclo crane	8	90 km/h	90 km/h	360	× 10 000

Looking at the transport capacities, expressed as multiples of "Man dragging logs" (the right column in table 2.3) the big difference between modes of transport are obvious. The capacity of a truck is about $3 \cdot 10^4$ times higher than that of a worker.

The development of transport techniques, apart from improving equipment, is largely achieved by the construction of roads for equipment of high capacity. Transport work is transferred to more efficient equipment as road nets are extended to penetrate the forest areas. But such roads carry a cost and at some stage a further extension is no longer economically feasible.

Cable systems are not included in table 2.3. The reason is that the aggregate time required to rig up and move the equipment to new settings and to "load" and "unload" are so high that a figure on the transport capacity in travelling along the line is not particularly significant and may easily be misunderstood.

Air born transport has attracted considerable interest in the past few decades. However, aircrafts and helicopters have a very high fuel consumption per unit transport work and technology is not likely to be able to change this fact.

Helicopters are likely to be feasible for a few special situations and for the transport of other goods than bulk commodities like timber.

Balloons are because of the size needed for sufficient lifting capacity very difficult to handle and control during severe weather conditions.

The cyclo-crane is one of the hybrids between helicopter and ballon developed in recent years. Such hybrids are designed to overcome some of the main shortcomings of each type.

River drive of loose logs is not listed in table 2.3. The transport routes have a very high capacity and consequently also low variable cost. The passage of hydroelectric power stations may require considerable input of labour and be more cost sensitive to quantity. River drive usually requires a costly sorting of logs on owners, species and assortment. If the mills are not located close to the assorting place, bundling and towing in barges is also required.

It should already at this stage be mentioned that roads have an important function for transport of labour and other material than timber. This will be further dealt with later on. The road net of a forest area is included in the now commonly used term infrastructure.

Assume that the cost of the timber truck is US$ 75 per hour the unit cost of transport is US$ 75 per hour divided by 1 033 t · km/h per hour or 0,075 US$ /t · km. This unit cost is the "variable direct cost" of this timber transport. It also follows that this cost can be decreased by increasing the speed, by increasing the load and by decreasing the operating cost per hour of the truck. All these three factors have the same influence. If the load is increased with 10% but leads to a reduction in speed with 10% the cost is not changed (provided the operating cost per hour of the truck remains the same).

In the table 2.4 an example is given of the variable direct cost of some common modes of transport. The same great variance can be found as earlier was shown for transport capacity.

Table 2.4 Example on variable transport costs for some different modes of log transport returning empty.

Means of transport	Capacity t · km/h	Cost per hour, US$	Variable trp cost, US$ per t · km
Man dragging	0,036	10	280
Horse, skidding	0,72	16	22
hauling	8,96	16	1,80
Tractor, skidding	11,52	30	2,60
forwarding	20,16	40	2,00
Logging truck	1 033	75	0,075
Helicopters	320-640	2 000-3 000	5,00-6,00
Cyclo Crane	360	1 500	4,20

Figures for helicopter and cyclo crane are computed from Guimier & Wellburn, 1984

As already stressed, the cost of the transport routes must also be considered (see Chapter 2.7) before anything can be said with regard to the economy of the transport mode.

2.4.3 Terminal Direct Cost of Transport. Before a transport unit starts out, the cargo must be loaded and when reaching the point of destination it is unloaded. There is an infinite variety of ways for loading and unloading involving more or less sophisticated equipment or elaborate terminals. The cost of loading or unloading is less useful to derive in the same manner as was made above for variable direct transport cost. The reason is that loading and unloading equipment is often used intermittently. It is standing idle for long periods and the cost per hour of effective work of such equipment will vary greatly with the ratio effective to idle time. As a rule of thumb one could possibly say that the cost of loading and unloading is seldom less US$ 0,40 and 0,20 per tonne respectively. Thus the direct terminal cost is of the same magnitude as the cost of moving the timber 20 km with a truck. It should be noted that the cost of the transport vehicle (e.g. the truck) when standing still for terminal work should be included in the direct terminal cost. (See further Chapter 3.6 Costing of Machines.)

Terminals are often designed to cater for other work operations than just the transfer of cargo from one mode of transport to another. In most cases the terminal has a buffering function to allow independency for the two links of transport meeting in the terminal. Buffering then serves the purpose to provide space for unloading timber or provide timber for immediate loading on the arrival of an empty vehicle. Transport systems designed for little buffering are often called hot logging, which always are more prone to disturbances in the flow of timber. For logging in steep terrain hot logging often is a necessity because of problems of finding storage space.

However, several other functions can be added and the manning of the terminal and machines or installations depends on what specific functions are included. Such functions can be

- scaling or weighting timber
- assorted mixed loads in market assortments
- crosscutting of long logs or stems
- debarking, chipping

2.4.4 Variable Indirect Cost - Cost of Transport Routes. All modes of transport require a road or a route. Even an aeroplane requires radio stations and similar installations for control and guidance which are more numerous the greater the distance. Only a man carrying a load and a homing pigeon with a message are examples of modes of transport without routes necessarily incurring work and costs. As a rule such routes or roads are used for transporting several types of goods or giving several types of services which should share the cost. Such cost sharing can only be worked out indirectly on the basis of some logic or principles - therefore the cost per unit of transported goods is of an indirect nature.

Assume that a truck road is built for the single purpose of hauling 100 000 tonne of timber from a forest to a mill over a distance of 40 km and that the cost of the road is US$ 200 000 (or 5 000 US$ per km) for construction and maintenance. The unit cost is then US$ 2 per tonne of timber and per unit of transport

$$\frac{200\ 000}{100\ 000 \cdot 40} = 0{,}05 \text{ US\$/t} \cdot \text{km}$$

It shall already now be repeated that such a road is always used for other movements and its cost should not entirely be charged on the timber.

The purpose of this example is to show that the road cost can be expressed in the same way as the actual movement of the transport goods provided we know the cost of the road during its lifetime and the volume expected to be moved on it during the same time. As the objective is to find a low aggregate cost of the transport system (the direct transport cost of the goods and the road cost), this derivation is of basic significance in the further elaborations in forestry transport economics.

2.4.5 Terminal Indirect Cost. For the loading and unloading of cargo certain installations are generally needed such as space for storage, loading platforms, ramps etc. Such installations are often used for a variety of goods and their costs must in one way or another be distributed on each specific product. In timber transport the following terminals are found:

Place	Terminology in use in North America	in the USSR
In the forest, e.g. wood piled along strip roads	Primary landing	-
Along truck roads	Secondary landing	Upper landing
At railheads, riverbanks or central conversion sites	Tertiary landing	Lower landing
At mill site	Mill wood yard	-

The terminals serve the purpose of quick, easy and cheap transfer of the goods from one mode of transport to another (or from one vehicle to another). The quantity to be handed per unit time or during the life time of the terminal determines how costly the installations shall be made and what equipment to use.

We shall here mainly deal with primary and secondary landings on which rather small quantities are handled. Consequently, the fixed installations are often kept to a minimum and portable loading machines are used which can easily be moved from one landing to the other. The better ability such loaders have to work on rough ground the lower is it possible to keep the terminal indirect cost.

There are some peculiarities with terminals in log transport which should briefly be mentioned already at this stage.

For a layman forest harvesting is simiply a matter of bringing logs from the forests to a mill site. In reality it is somewhat more complex. To make the best use of these logs they must be assorted in groups according to a variety of criteria stemming from the ultimate use of them. Two such criteria are always present, tree species and diameter of the log, but others are often also involved, e.g. peeled and unpeeled. As from the beginning all logs are mixed, they must be sorted at some stage during the transport. This sorting is generally made on the landings. What is of interest here is that the sorting of mixed logs in 2 or possibly 3 groups can often be made very cheaply if this sorting is made in conjunction with a loading or unloading operation. A crane loading or unloading a tractor can often very easily and without any sizable cost increase separate the logs in two groups. But as soon as there are more than 3 groups such combined loading and sorting is becoming increasingly difficult and one must often resort to splitting the operation into two - one loading and one sorting operation. The more groups of logs the more spacious must the landing be for the same total volume. In fact, the need to sort the logs often poses rather tricky planning problems in organizing the transport and must often be given careful consideration.

Scaling - to assess the quantity and often also the quality of the logs - is often made on landings often adding to difficulties of planning terminals.

Lastly, it should be mentioned that it is often very difficult to provide landings of sufficient size along truck roads in steep terrain. This difficulty may often determine how the whole transport system shall be organized and what equipment should be used.

2.5 ON SPACE AND TIME

Space and time is the reference frame of our experiences and observations from the world we live in. Only after identifying the significance of these dimensions, can we arrange our knowledge meaningfully, remember and use it intelligently and constructively.

These dimensions of space and time have particular significance for forestry as they can be used to demonstrate very instructively some of its major problems to which they are closely related. Many of the readers of this text may in earlier school years have worked with problems related to space - geometry - and innumerable such problems have with more or less headache been solved. Some may remember these exercises as a stimulating intellectual play - others as a refined torture with abstract figures and lines staring from a piece of paper, many times only too blank. It is often easily overlooked that later in life, it is exactly the same lines and distances we use in transport studies, that those areas where we plant, tend and harvest the trees are subject to the same geometrical laws as our school triangles, squares and polygons and that logs and loads of wood are stereometric figures just as the pyramid or the frustum of a cone. Thus, the step from the abstract intellectual world of the school to learning for real life problems does not exist when it comes to theories and principles but it may many times be somewhat difficult to take simply because of lacking training in identifying the physical world before our eyes with abstract scaled-down models.

The perception of real life situations is greatly facilitated by insights and knowledge through which the nature and structure of various phenomena are understood. Such an analytic approach involves the transformation of the physical world into a homomorph, abstract picture - a model - to be manipulated and scrutinized in various ways for the purpose of exploring its behaviour under various sets of actions.

2.5.1 <u>Distance</u>. Industrialization and the gradual development of market economics has vastly increased the transport of raw materials and goodsboth in quantities and in distance. The component of transport in the final price of the products has tended to increase. This cost increase has been offset to a degree by the lower processing costs that large scale factories and markets offer.

The shortest distance between two points - say between the places of production and of consumption - is the straight line. But in modern economies the transport flow of goods often follows rather complex patterns and wandering routes from place of origin to the scrap pile. A piece of log grown in the tropical forests of Indonesia may very well have travelled in log form across the Pacific to a veneer mill on the US West Cost and end up in a door panel in some office in Djakarta. Forests products follow the same general transport model as most other goods - tree mirrored in the water with a processing plant at the base of the tree. The plant is the centre for collecting the raw materials and for distribution of the products.

However, the distance of transport has a direct impact on the price of the product and this is particularly so for products of low value per unit of weight or volume. Forest products are exactly such goods which also are typified by a rather low value added in the processing forest industries. It follows that it is of fundamental importance for forestry to master the cost increases incurred by distances through an efficient transport system both for bringing the timber from the forests to the mills and for distributing the products to the consumer. The location of the mills and their capacities must be carefully considered with the regard both to the supply of timber and the markets for the products. In principle the best location is one minimizing the aggregate transport cost, which many times may be the same as minimizing transport work or average distances of transport for raw materials and products. Another important aspect is the economy of scale of the forest industries, which is particularly evident for pulpmills but less important for sawmills and other mechanical wood industries. The decrease in production cost by increasing size of a pulp mill may often offset the higher transport cost of pulpwood, which then must be drawn from a larger area. The problems of distribution and marketing of processed forest products will not be included in this book. One of the reasons is that these problems, in contrast to forest transports, are strongly influenced also by other factors outside the scope of this text.

A certain consensus is useful for the terminology on distances of transport, e.g.

(i) <u>straight line distance</u> (or "as the crow flies"), in ground plane

(ii) <u>(real) transport distance</u>, which is the actual road travelled with payload

(iii) <u>winding factor</u> $= \frac{\text{real transport distance}}{\text{straight line distance}}$

always >1 and often in the magnitude of 1,3-1,6 for forest transports. The winding factor includes the lengthening incurred both by horisontal and vertical curves as well as for difference in altitudes.

Difference in altitude can often be ignored for long hauls and need only be considered for logging in steep terrain, where the straight line distance often is not stated in the ground plane but along the slope.

Logging is almost always an one-direction transport. The payload is carried only in one direction and the carrier returns empty. In detailed studies it is often necessary to distinguish between the transport distance of the carrier and that of the cargo. In the extraction of logs by tractors, the logs or boles are often picked up along a strip road and there are three phases of the roundtrip:

(i) travelling empty
(ii) travelling between the loading stations, partly loaded
(iii) travelling fully loaded

In such cases it is logical to assess the transport distance as the weighted average of the travelled distances of the logs. The average transport distance of the cargo is in such cases smaller than the average roundtrip distance of the carrier and for some techniques this difference is considerable. For further details see Appendix A to Chapter 2.

2.5.2 Area. Forestry is a production which like agriculture takes place on an area. This distinguishes them from many other branches of the economy for which the dependence on an area is less apparent. Many of the forest work operations take place on and in the area and many of the difficult or costly problems of forestry derives from this "area dependency". This is the reason why a recapitulation of some basic concepts in geometry are useful in the analysis of forest work operations.

Already the old Greek mathematicians proved that an area can be divided (and fully covered) only by certain uniform polygons, the triangle, the square and the hexagonal. The pattern of these three polygons are shown in the figure 2.1. A fourth case is shown in figure 2.2 where the area is divided by equally spaced, parallel lines - in principle rectangles with two infinite sides. A forest area can of course be divided in many other ways - but not by other uniform polygons - but such divisions will always fall somewhere in between these four cases from the point of view of this scrutiny.

We shall now study some of the main implications of the area dependency, starting with a recapitulation of basic geometry. Next, modifications of these geometrical figures will be made by introducing correction factors so that a better conformity with real life situations can be achieved. This adaptation to reality is necessary as the topography of forest areas usually does not allow optimal geometrical configurations of the technical arrangements, like forest road nets. Also, the work operations may require deviations from work procedures which minimize transport distances and movements of timber, workers and machines.

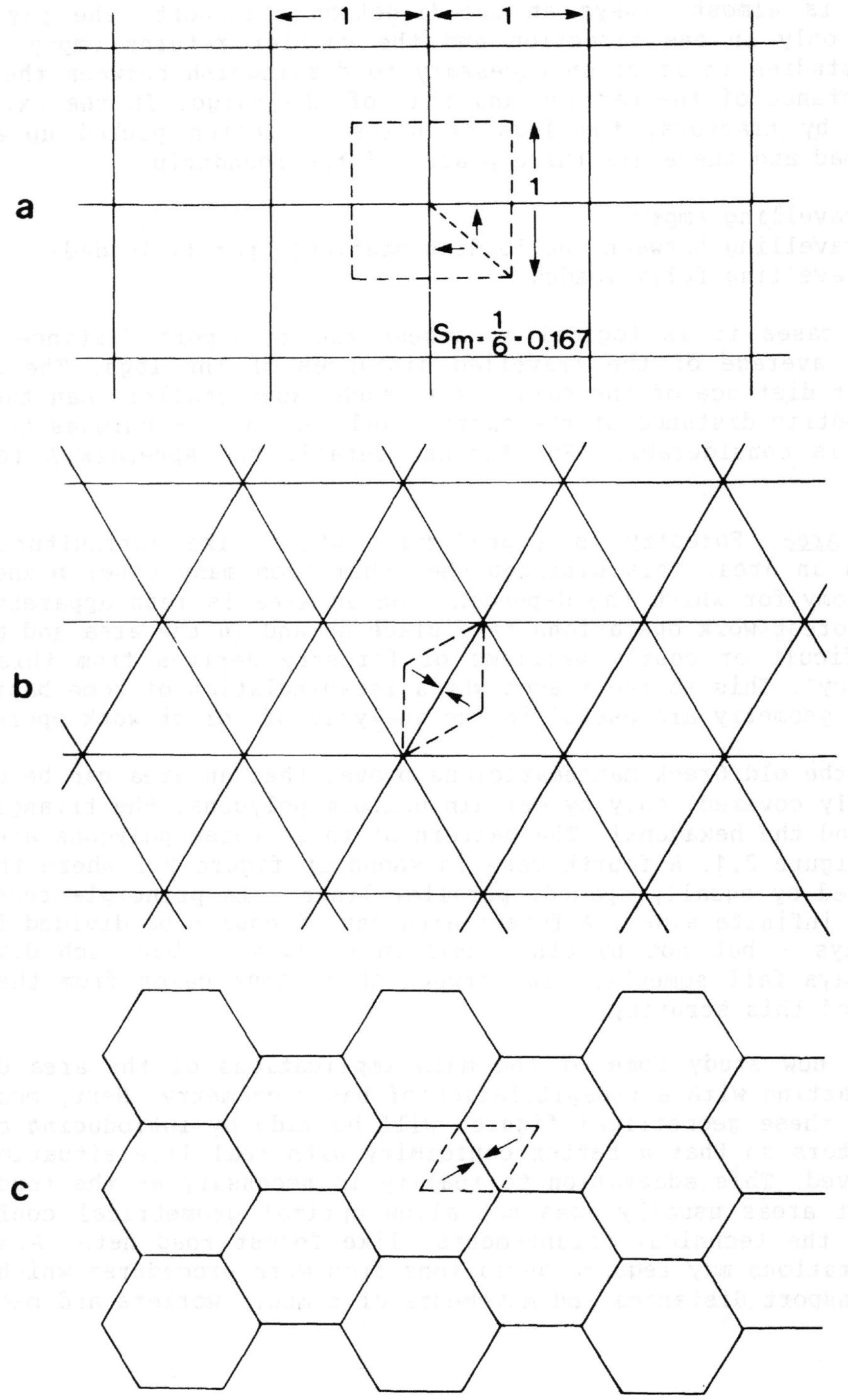

Figure 2.1. Polygons giving full coverage of an area
a = squares
b = triangles
c = hexagons

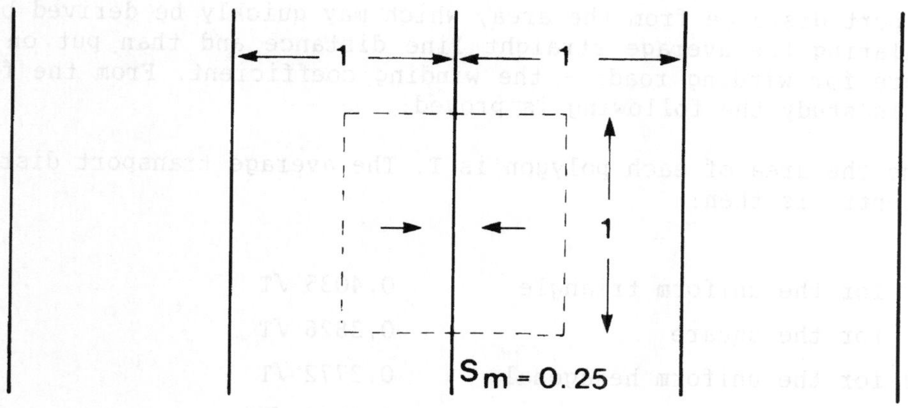

Figure 2.2. Area divided by infinite, parallel roads.
S_m = average transport distance for the area.

We can now let these figures be models of some technical arrangements in our forests. We start by letting the sides of the polygons be roads to which the logs shall be extracted or, inversely, from which some material shall be brought out on the area, e.g. plants or fertilizers. We are now interested to know the average distance from the area to these roads, for forest problems thus assuming a uniform density of logs, plants, fertilizer etc over the area. In order to make a comparison of these road models meaningful the same road density is assumed for all four cases, namely, 1 unit length of road per 1 unit area. In fact the figures on the squares, triangles and hexagonals are drawn in scales so that the road density in the three cases is the same.

In the case with infinite parallel roads (figure 2.2) it is evident that the average transport distance from the area is 1/4 of the spacing between the roads. In the three other cases - the triangles, the squares, the hexagonals - it can be proved that the average transport distance will increase with one third, or 33%, assuming equal road density. This can also be expressed so that the efficiency for "draining" an area of road nets forming uniform triangles, squares or hexagonals is only 75% (1:1.33) of the most efficient one, that with infinite parallel roads. The efficiency of road nets forming uniform, but finite, rectangels will be between 75% and 100% depending on the spacing between the cross roads (= the short sides of the rectangles). However, forest road nets are always much more irregular than those so far exemplified. A method of measuring the efficiency of real road nets was developed by von Segebaden (1964). In Appendix A to Chapter 2 a full account is presented on the "geometry of forestry".

Up to now the sides of our polygons have symbolized roads. Let now these sides symbolize boarder lines of area figures for the purpose of studying the average transport distance of these figures to the centre of these polygons. Such a case can be visualized by placing a sawmill in the centre of a polygon. The saw mill manager wants to know the transport distance from the area, which may quickly be derived by first calculating the average straight line distance and than put on an allowance for winding roads - the winding coefficient. From the findings of this study the following is proved:

Assume the area of each polygon is T. The average transport distance to its centre is then:

(i)	for the uniform triangle	$0,4035 \sqrt{T}$
(ii)	for the square	$0,3826 \sqrt{T}$
(iii)	for the uniform hexagonal	$0,3772 \sqrt{T}$
(and	for the circle	$0,3761 \sqrt{T}$)

From this we can conclude that the division of a forest area in uniform hexagonals is the most efficient division if we want to minimize the transport distances to the centres of the respective area figures. The difference between the square and the hexagonal is rather small. For practical purposes, the square pattern can be used in order to facilitate the calculations. The circle shall be excluded as circles cannot give full coverage of the total area. We can express these average distances on the basis of the distance between the centres of the polygons, "a". The average transport distance then takes the following values:

(i)	for the uniform triangle (a = 2/3 of its height)	$0,4599 \cdot a_{triangle}$
(ii)	for the square (a = its side)	$0,3826 \cdot a_{square}$
(iii)	for the hexagonal	$0,3510 \cdot a_{hexagonal}$
		or 0,6080 of its side
(for	the circle with the radius 1	0,333)

Assume that sawmills of equal capacities in an area spaced 50 kilometers apart and have divided the supply area in a hexagonal pattern. The average transport distance (straight line) is then $0,3510 \cdot 50 = 18$ km. Assume a winding coefficient of 1,4 and the real average transport distance becomes $18 \cdot 1,4$ or about 25 km.

We have up to now studied road nets and transport distances for some very stereotype and homogeneous cases. More realistic road nets will be further elaborated in the chapter 3.5. But before leaving the subject of average transport distances we shall drop the assumption of uniformity and homogenity and explain suitable approaches for assessing average transport distances from an area in more realistic situations.

If the density of the transport goods (say timber, plants, fertilizer etc) varies over the area, one need only to subdivide the area in sub units, assess each sub unit separately and by weighting calculate the total average by using the formula

$$s_{ave} = \frac{\Sigma\, q_i \cdot s_i}{\Sigma\, q_i} \qquad (2.2)$$

where q_i = quantity of transport goods for the area unit "i"

s_i = average transport distance for the area unit "i"

Transport from an area to a road (and also transports in the opposite direction) usually terminates (or starts) at certain landings. Logs are usually not piled up uniformly along the roads. Up to now it has been assumed that each movement of the goods was to (or from) the nearest point of the road but this is seldom true - a concentration to certain points or landings along the road is a more common pattern. Two methods for dealing with such situations are suggested. The first is to simplify the area to one or several rectangles and select the appropriate factor in table 2.5. depending on the location of the landing. This table in turn is based on a formula for calculating the average transport distance for a rectangle to one of its corners as explained in Appendix A to Chapter 2.

Table 2.5 Average transport distance at different ratios short/long side of a rectangle at different location of the landing. Distance as a ratio of the long side. The column numbers correspond to the location numbers in the rectangle, figure 2.3.

Ratio short to long side	Average transport distance as a fraction of the long side with the landing located. in corner	at the middle of short side	middle of long side	on the line from the middle of the short side at a distance of 1/4 of the long side	1/3 of the long side	1/2 of the long side	the long side at a distance of 1/4 of the short side	1/3 of the short side	1/2 of the short side
(1)	(2)	(3)	(4)	(5)	(6)	(7)	(8)	(9)	(10)
0,05	0,502	0,501	0,253	0,313	0,279	0,251	0,252	0,251	0,251
0,1	0,506	0,502	0,260	0,316	0,281	0,253	0,255	0,254	0,253
0,2	0,521	0,506	0,283	0,322	0,288	0,260	0,266	0,263	0,260
0,3	0,541	0,513	0,312	0,332	0,298	0,271	0,281	0,275	0,271
0,4	0,565	0,521	0,346	0,343	0,310	0,283	0,299	0,290	0,283
0,5	0,593	0,530	0,383	0,357	0,323	0,297	0,319	0,307	0,297
0,6	0,624	0,541	0,422	0,371	0,338	0,312	0,340	0,325	0,312
0,7	0,657	0,553	0,463	0,386	0,354	0,328	0,363	0,344	0,328
0,8	0,691	0,565	0,505	0,402	0,371	0,346	0,387	0,364	0,346
0,9	0,728	0,579	0,549	0,419	0,389	0,364	0,412	0,385	0,364
1,0	0,765	0,593	0,593	0,437	0,407	0,383	0,437	0,407	0,383

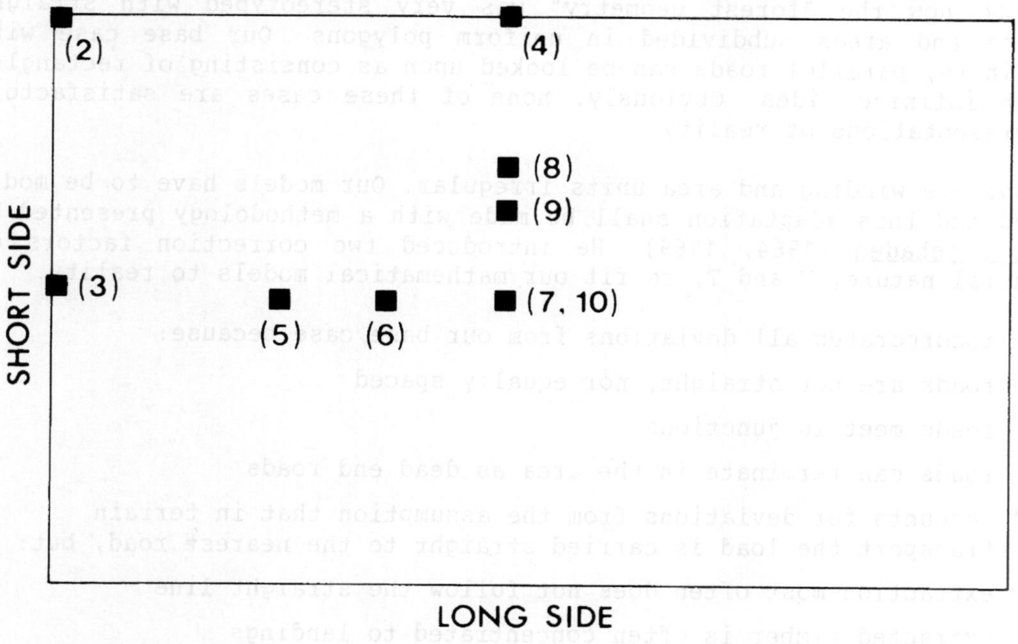

Figure 2.3. Locations of landings, corresponding to the column numbers in the table 2.5.

The second method is based on a division of the area in a large number of sectors of a circle with its origin placed on the landing. Two formulas are available

$$S = \frac{2}{3} \cdot \frac{\Sigma r_i^3}{\Sigma r_i^2} \qquad (2.3)$$

in the case the sectors have the same angle, or

$$S = \frac{2}{3} \cdot \frac{\Sigma r_i}{N} \qquad (2.4)$$

in the case the sectors have the same area.

Where r_i is the radius "i" of a sector crossing the perimeter of the area and N is the number of sectors. The first formula is the most practical one. See further Appendix A to Chapter 2.

Up to now the "forest geometry" was very stereotyped with straight lines and areas subdivided in uniform polygons. Our base case with infinite, parallel roads can be looked upon as consisting of rectangles with infinite sides. Obviously, none of these cases are satisfactory representations of reality.

Roads are winding and area units irregular. Our models have to be modified and this adaptation shall be made with a methodology presented by von Segebaden (1964, 1969). He introduced two correction factors of spatial nature, V and T, to fit our mathematical models to reality.

"V" incorporates all deviations from our base case because:

roads are not straight, nor equally spaced

roads meet in junctions

roads can terminate in the area as dead end roads

"T" accounts for deviations from the assumption that in terrain transport the load is carried straight to the nearest road, but:

extraction most often does not follow the straight line

extracted timber is often concentrated to landings

extraction may sometimes not go to the nearest road.

The numeraire of "V" can now be established for real road networks by putting a grid of points - preferably randomized - over a map of the road system, measure the distance from each point to the nearest road (s_i) and form

$$\bar{s}_r = \frac{\Sigma s_i}{n} \tag{2.5}$$

Index "r" denotes real road network "n" number of measurements

and $\bar{s}$ average distance.

$$V = \frac{\bar{s}_r}{\bar{s}_b} \tag{2.6}$$

Index "b" denotes the base case.

We already showed that in the case of road nets forming triangles, squares and hexagonals "V" had the value of 1.33. The studies of von Segebaden, Bachmund (1966), Samset (1975) and Abegg (1975) can be summarized as follows:

For very well planned road nets, easy terrain, many dead end roads $V = 1,25 - 1,35$

For road nets in normal terrain, few dead end roads $V = 1,35 - 1,50$

For road nets in difficult and very difficult terrain $V = > 1,50$

$s_{b,r}$ shall as a rule always be measured in the horisontal plane.

As an interesting comparison it can be noted, that if we have a grid of straight lines, randomly spread out over an area, "V" will equal 2, according to a proof by the French mathematician Buffon. We shall make use of this proof later on in the text when dealing with estimates of road densities in inventories.

In steep or very steep terrain the transport to and from the roads may for technical reasons take place in one direction only. Then, in our base case

$$\bar{s}_b = \frac{RS}{2} \qquad (2.7)$$

RS = road spacing

and this ratio should then be applied in subsequent elaboration of the spatial geometry.

The numeraire of the "T" factor is established on empirical measurements of straight line distances, respective travelled distances (e.g. along winding roads) by measurements;

(i) on maps from a grid of points
(ii) from random points in the terrain
(iii) in work study by recording travelled distances for each log, stem or tree, assuming they are evenly distributed over the area.

$$T = \frac{\bar{s}_r}{\bar{s}_b} \qquad (2.8)$$

"T" often takes the value 1.2 - 1.5 for increasingly difficult terrain.

In our base case with maximum road efficiency (infinite, parallel roads) we proved that

$$\bar{s}_b = \frac{RS}{4} \qquad (2.9)$$

We shall now introduce the concept "Road Density" (RD) = length of road per unit area. The average extraction distance is inversly correlated to RD and for our base case:

$$\bar{s}_b = \frac{1}{4\ RD} \qquad (2.10)$$

RD = road density

In order to go over to real, average transport distance we shall apply the "V" and "T" factors.

$$\bar{s}_r = \frac{V \cdot T}{4\ RD} \quad \text{or } RD = \frac{V \cdot T}{4\ s_r} \qquad (2.11)$$

Taking rather normal values V = 1.4 and T = 1.3 for rather good terrain

$$\bar{s}_r = \frac{0{,}455}{RD} \qquad (2.12)$$

Example: Having a road density of 2 km km^{-2} (or 2 miles per square mile)

$$\bar{s}_r = \frac{0{,}455}{2} = 0{,}2275 \text{ km (or mile)}$$

In countries on the metric system the road density is often in practise assessed in meter per hectare, which is easily visualized but not mathematically correct. Extraction distances are often given in km. We can then form what we may call a "Spatial Constant" (SK).

$$SK = V \cdot T \cdot 2{,}5 \qquad (2.13)$$

Using the results from the studies, previously referred to

SK equals 4 to 5 for flat and undulating terrain
5 to 7 for hilly terrain
7 to 9 for steep terrain
9 or more for very steep terrain

The spatial constant thus goes from 4 to 9 or more. If we for example work in moderate terrain situations we can use the rule of thumb that 5 divided by the road density (in meter per hectare) gives the average extraction distance in km. Or if the average extraction distance is known we can reverse the procedure and derive the corresponding road density.

Von Segebaden also found that there is for forest road nets in Sweden a rather stable distribution of the area on distance zones from the roads as shown in figure 2.4. If for example the average extraction distance is 200 metre one could expect that the roundtrips from that operation are distributed as follows:

60% of the roundtrips will average around	100 metre	(-200 metre)
30% "-	300 "	(200-400 ")
7% "-	500 "	(400-600 ")
3% "-	700 "	(600-800 ")

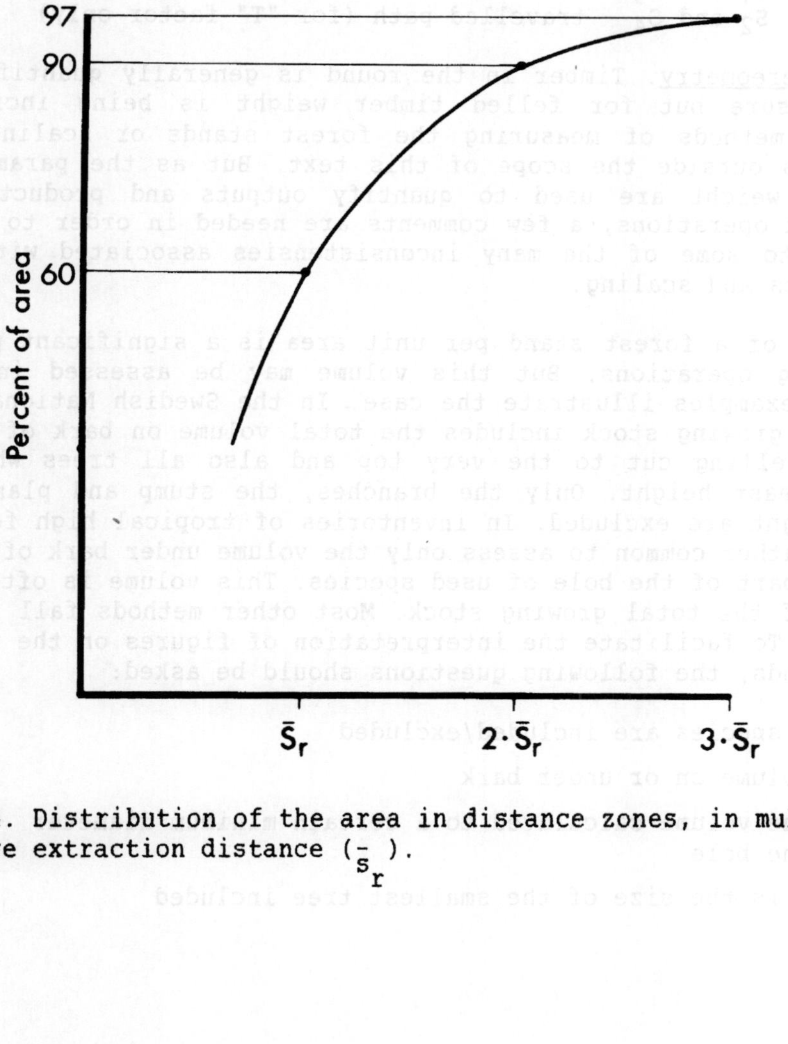

Figure 2.4. Distribution of the area in distance zones, in multiples of the average extraction distance ($\bar{s}_r$).

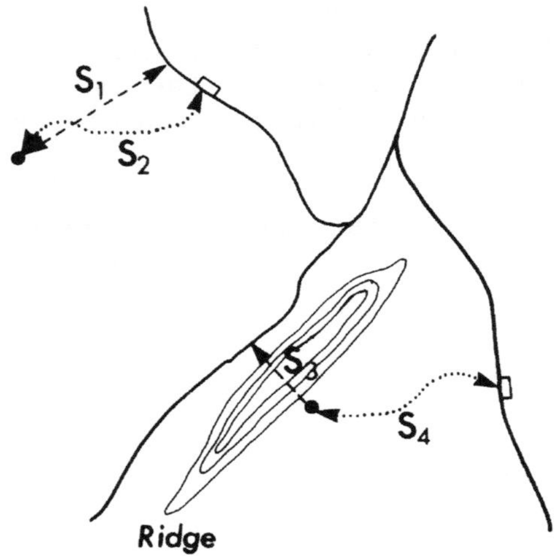

Figure 2.5. Measurements of distances from random points for establishing "V" and "T" factors.
S_1 and S_3 - straight line distances to the nearest road
S_2 and S_4 - travelled path (for "T" factor only)

2.5.3 Stereometry. Timber in the round is generally quantified in a volume measure but for felled timber weight is being increasingly used. The methods of measuring the forest stands or scaling of the products is outside the scope of this text. But as the parameters of volume or weight are used to quantify outputs and productivity in forest work operations, a few comments are needed in order to draw the attention to some of the many inconsistensies associated with timber measurements and scaling.

The volume of a forest stand per unit area is a significant parameter for logging operations. But this volume may be assessed in various ways. Two examples illustrate the case. In the Swedish National Forest Survey the growing stock includes the total volume on bark of the bole from the felling cut to the very top and also all trees which have reached breast height. Only the branches, the stump and plants below breast height are excluded. In inventories of tropical high forests it is still rather common to assess only the volume under bark of the merchantable part of the bole of used species. This volume is often only a fraction of the total growing stock. Most other methods fall inbetween these two. To facilitate the interpretation of figures on the volume of forest stands, the following questions should be asked:

(i) what species are included/excluded

(ii) is volume on or under bark

(iii) is the volume calculated to a certain minimum diameter of the bole

(iv) what is the size of the smallest tree included

As to the determination of the volume of logs, the scope in efficiency studies should always be to state the quantity in true volume, on or under bark. But logs are most often scaled for marketing or utilization purposes and seldom for the purpose of measuring logging efficiency. There exist a vast number of scaling systems adapted to local traditions and situations. They have only one thing in common - they never give the true volume, possibly with a few exceptions. Many of them are obsolete as both the timber and the utilization techniques have changed drasticly. Deviations from the true volume have two origins. One is related to scaling techniques, the other to computing methods and both aspects should be considered in attempts to convert into true volume. Only a few, rather randomly selected, causes of error shall be mentioned as being of special interest in logging studies.

(i) How are cull trees, cull logs or volume defects of the log assessed.

(ii) Is trimming allowance included or excluded.

(iii) At what point of the transport is scaling carried our and what are the eventual losses before or after scaling.

In order to compare log volumes of different scaling systems conversion factors are worked out but often such conversion factors are imperfect. One example is the conversion of the many local "board feet" scales in North America, for which conversion to true volume with the aid of a standard conversion factor rarely can be made with accuracy, especially not for smaller quantities of timber. (Conversion factors are given in a table in "Units and Conversion Factors" in the last pages of this volume.)

The conversion of logs or chips that have been weighed into true volume is often made by taking samples for the determination of conversion factors. This is generally recommendable as the moisture content may vary considerably with length and conditions of storage. The ratio sapwood to heartwood for the same specie may also incur bias. Saw mill chips, the bulk of which originates from the sapwood portion of the logs often have different weights than chips from pulpwood in the round and this often applies both to the actual green weight and to bone dry weight.

The problem of measurement of timber and wood fibre must consequently be observed both in comparisons of production in logging and in planning logging studies. One example illustrates the latter case. In logging studies volume is often determined by measuring the length and some diameter of the logs. It is standard practise that the logs are grouped in diameter classes and the volume for each such class computed, often in volume tables. It may now be of importance how such a volume table originally was computed. Some tables are made on the basis of the lowest value of the diameter class, others on the basis of the mid-diameter of the class. The relevant diameter will of course depend on the dispersion of the population of logs within the class and this will vary from case to case. If logs are scaled in centimetre classes, the difference in volume between "low" or mid class will for the 20 centimetre class be:

$$\frac{20,5^2}{20,0^2} = 1,05 \text{ or } 5\%$$

But for the 5 centimetre class the difference will be

$$\frac{5,5^2}{5,0^2} = 1,21 \text{ or } 21\%$$

The error of this type will of course depend on the width of the class in relation to its absolute value. In work studies it is therefore advisable to scale logs to the nearest class value, e.g. all logs between 19,51 and 20,49 centimetre are tallied in the 20 centimeter class.

The diameter of a log or a bole bears strongly on the production rate in many logging operations. It is therefore useful to explore some stereometric properties of a log, tapering from butt to top as illustrated in figure 2.6. We find that

(i) the volume per unit length is proportional to the square of the diameter
(ii) the surface of the log per unit length is proportional to the diameter
(iii) surface of the log per unit volume is inversely proportional to the diameter.

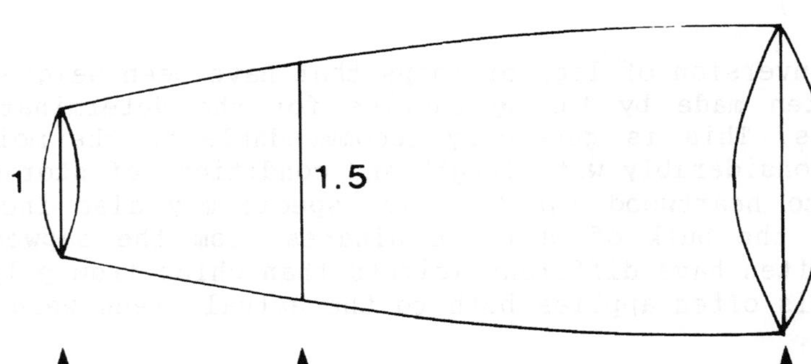

	1	1.5	2
Volume per unit of length	0.79	1.96	3.14
Log surface per unit of length	3.14	4.71	6.28
Log surface per unit of volume	4.0	2.4	2.0

Figure 2.6. Stereometric properties of a log

Let us match this properties with the "mechanics" of some work operations:

Manual debarking: Assuming, bark properties uniform, a man debarks a constant area per unit of time (except when logs are very small). Consequently his output is in direct proportion to the log diameter, e.g. double volume output for 20 cm pulpwood sticks as compared to 10 cm sticks.

Rotary ring debarkers: These machines have generally a pre set feed rate - logs are fed into the machine longitudinally at a constant speed. The volume output is therefore proportional to the square of the log diameter, e.g. fourtimes the volume output for 20 cm logs as compared to 10 cm logs.

Cutterhead barkers (rossers): Usually with cutterheads removing the bark from a log being rotated and simultanously fed forward, have in principle the same ability as a worker - to remove a constant surface of bark from the log per unit of time. Consequently volume output is directly proportional to the log diameter.

Drum barkers: Capacity generally independent of log diameter.

This comparison somewhat simplifies reality but nevertheless it throws light on one of the fundamental facts on logging operations - the correlation between log size and work output. But as we have seen, this correlation takes different forms depending on the "mechanics" of the work. The following generalizations can be made.

Output correlated to D^2: Example above debarking with rotary ring debarkers. Also valid for all operations where a worker while performing proceeds along the bole - such as debranching of felled trees of logs with axe or power saw - or where the boles or logs while being worked at travel at constant speed through a machine, e.g. a wood chipper or a gang saw, also all handling of logs piece by piece in cases of a constant cycle time. But the correlation may sometimes be modified by technical arrangements. On rotary ring debarkers or gang saws the feed rate can automatically or manually be varied with log diameter with the purpose of tempering the sharp decline in output with decreasing diameter. But it is obviously rather easy to set up a mathematical expression for output as related to log diameter once we know:

(1) that the correlation is basicly Output = a constant $\cdot D^2$

(2) the modification incurred by increasing feed rate with decreasing diameter.

In doing so we have established a mathematical function - a model - for the work operation and such models are extremely useful in studies to plan for or improve work efficiency and economy. It shall only be pointed out here that a clear insight in such basic correlations and principles are of great help in the analysis of work operations and in drawing conclusions and in making choices and plans for them.

Output correlated to D: All work operations where a constant area of the log surface is worked per unit time, such as manual or rosser debarking. Another example is the peeling of veneer from logs rotating at constant RPM.

Output independent of D: Drum barkers and all operations where trees or logs are handled in bundles of fixed size or weight.

2.5.4 Time. Our fourth dimension, time, has for our purpose three aspects:

(1) Time is a scarcity factor in the sense that we will have to pay a price for each unit of time we make use of an asset (but each time unit during a day, year or decade may not necessarily have the same price). An asset is used to produce some goods or service and the more it produces per unit of time, together with other assets, the lower will the cost of this asset be per unit of production. Consequently, measurement of time is needed in most efficiency studies and the efficiency is, as will be found on many occasions in this book, often quantified as output per unit time.

(2) Each asset has a present value, the value a buyer puts on it and is willing to pay today to get it immediately. If the asset is not immediatly available but at some future point in time, the value or price the buyer puts on it is less. This decrease in value of price is an expression of the buyer's time preference. It is standard practise to state the decrease in value for a period of one year (but any period of time is in principle just as usable) and to quantify the decrease in hundredth per unit value. It is called the rate of interest to signify how quickly the buyer losses interest the further away in a distant future he can lay hands on the asset he wants. The application of rate of interest offers possibilities to rank the present value of various assets being available at different points of time in future, or of a series of assets becoming available at regular or irregular intervals. To facilitate such assessments tables are computed on the present value of future events for different periods and rates of interest. It shall be observed that the use of such tables and rate of interest is not limited to money or capital. Any quantity being available at some point in time has a similar present "value". Assume that a forest property produces 4 cubic meter per hectare. The present value of this production at 5% rate of interest is 20 x 4 = 80 cubic meter (as the capitalized value for infinite payments of 1 at 5% rate of interest is 20). Assume a buyer estimates a profit of 7 US$ per cubic meter the present value of the forest is 7 x 80 = 560 US$ per hectare. But valuation of forest properties is not always as simple as that.

(3) The passage of time always involves change. If we lived in a world that did not change, next year would be exactly the same as this year. Forecasts and planning would be unnecessary as it would just be a matter of repeating what was previously done. But the dynamic development of our world, which is a function of time, requires abilities, skills and intellectual efforts to foresee future events in order to conform and adapt our activities so that they meet our expectations even in a changed situation. Perhaps still more important is the awareness of change and a recognition that nobody can predict the future even with rather modest requirements on accuracy. The advantage of retaining as high degree of flexibility in planning for future is obvious important to allow for successive adjustments to situation changes.

In economics, a distinction is generally made between using funds for investment and for consumption. This distinction is an arbitrary one as it is common practise to classify all fund-spending of assets used up within one year as consumption but investment if it takes more time then one year to use up the purchased asset. This convention is, of course, based on the fact that one year is the common time period for accountability of an enterprise and also for fiscal and tax purposes.

2.6 TRANSPORT SYSTEMS

2.6.1 Introduction. An industrialized society is largely characterized by a division of work, by specialization and concentration of the manufacturing of certain products in large industrial plants. The underlying reason for this is that by specialization and concentration the various factors of production - such as labour and equipment - are brought together under work conditions offering options of efficiency for a high level of production. Such conditions are created by a qualified staff for managements, planning and supervision of the work, by the use of equipment and machines and as a rule also by a higher input of energy such as mechanical power. The resulting higher output per unit of input - higher productivity - is often called the economy of scale.

However, the economy of scale in the processing of products must be bought at a price for longer transports. Raw materials, semi-finished goods and often also labour and energy must be moved longer distances. The same often holds true also for the marketing of the products. The cost component of transport progressively increases the more industrialized the society becomes. However, this increase is counteracted by more efficient transport system. Such systems are both a prerequisite and a consequence of industrialization. Developments in manufacturing and transport must go hand in hand and should preferably be integral parts in developments.

The relation between economy of scale and the transport cost is well illustrated by the pulp industry. The cost of production decreases with the size of the pulpmill but the larger the mill the longer distances must the pulpwood be brought. It is obvious that there is a limit of size of the pulpmill over which the cost increases for longer transports exceeds the sunk cost of pulp production. To establish exactly the pulpmill size with the lowest aggregate cost of pulping and pulpwood transport must be a problem of considerable practical interest. This textbook is intended to explore analytic methods by which such problems of optimization can be tackled and solved.

Sawmills are smaller and more numerous than pulpmills. Why is it so? The main reason is that the value added in making lumber of sawlogs is considerably lower than in making pulp. Higher capacity in a sawmill therefore offers smaller opportunities to decrease the cost in absolute terms whereas the counteracting factors are the same ones. Another reason is that with an increase in size there follows the need for more capital, not only for buildings and machines but also operating capital. Such money is much more difficult to borrow and therefore more expensive (higher rate of interest). As sawmills largely are generally owned by smaller firms the capital needed for expansion, especially the operating capital, limits the incentive to expand.

In chapter 2.3 it was pointed out that the term <u>transport</u> in a wide sense connotes all the activities <u>directly</u> or <u>indirectly</u> related to the movement of goods, people and materials. These activities as well as the physical assets such as roads, vehicles, terminals etc form a <u>transport system.</u> What are now the characteristics and main features of a forest transport system?

Firstly, the system is not only vehicles and labour. It also includes roads, terminals and other similar installations. Secondly, the system shall move not only logs but also labour and materials. It shall also provide services necessary for the growing and protection of the forest and also serve other uses of the forests, such as tourism, access for hunters and fishermen etc.

In chapter 2.7 the area dependency of forestry is outlined. This leads to widely varying quantities of transport in the different links of the transport from the stump to the mill, which in turn calls for different modes of transports. The task is now to design a transport system to perform the movements at lowest possible aggregate cost. To do this we need to know the capacities, cost and technological features of all the modes of transport, that are available.

2.6.2 <u>Modes of Transport at Varying Road Costs.</u> In the tables 2.3 and 2.4 the most common modes of transport employed in forestry are listed with regard to capacity, operating cost per hour and variable transport cost. There are very great cost differences as shown by the cost ratios in the table 2.6 for the most common modes of forest transport.

As the volume of transport successively increases downstreams in the system, these will be points in the area at which a change of transport modes shall be made. These points shall be identified.

Table 2.6 Approximate ratios of variable (distance) transport cost per $t \cdot km$ for some modes of forest transports. Truck haul on main road = 1.

	Ratio
Truck haul on main road	1
Tractor in extraction work	30
Animals in skidding	300
Man dragging logs in the terrain	4 000

Each transport requires a road, the cost of which shall be included when a choice is made. In general, better and more expensive roads are provided the higher volume of traffic that is expected for the period of planning of the transport system. The correlation between total cost at varying traffic volume is illustrated in figure 2.7 for some modes of transport. The graph shows a number of break-even points denoting that there always is a best mode of transport for a given traffic volume, which gives lower total cost than any other option. Each line in the graph has the form

$$TC = RC + Q \cdot vdc \quad \text{where} \qquad (2.14)$$

TC = total cost
RC = road cost
Q = traffic volume
vdc = variable direct transport cost per unit of traffic volume

The break-even point between two modes of transport occurs for a Q that gives the same TC for both, or mathematically

$$TC_1 = TC_2 \quad \text{or} \qquad (2.15)$$

$$RC_1 + Q \cdot vdc_1 = RC_2 + Q \cdot vdc_2 \qquad (2.16)$$

The break-even point is found if we solve Q in this formula

$$Q = \frac{RC_1 - RC_2}{vdc_2 - vdc_1} \qquad (2.17)$$

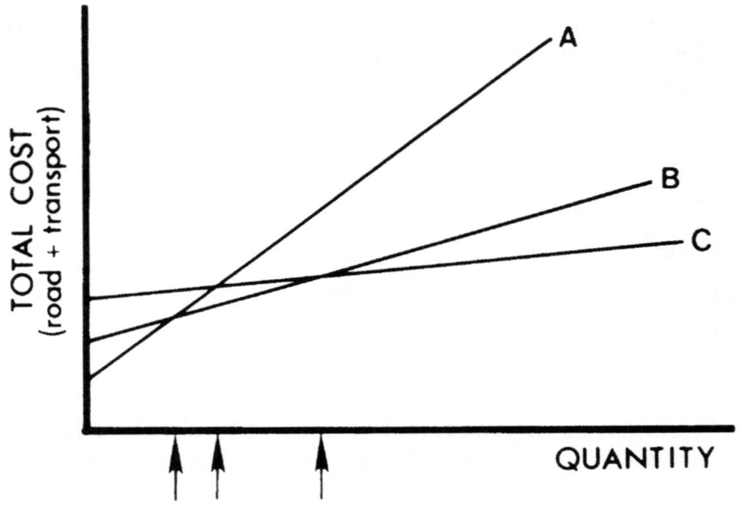

Figure 2.7. Graph illustrating the breakeven points (in quantity of transport) for three modes of transport A, B and C.

A characterized by low road cost, high variable transport cost
C characterized by hig road cost, low variable transport cost
B intermediate
Break even points indicated with arrows.

This relationship does not only apply to different modes of transport but also to the choice of road standard (road class) for one transport form.

Let us assume that four standards of forest roads are used. Road standards are often described in terms of travelling speed of a logging truck and the variable transport cost is proportional to the inverse of the speed.

We now want to know at which traffic volume one should move from one road class to another. This break even point is established by the formula (2.17). For road classes I and II the break even point is

$$Q_{I/II} = \frac{5\ 000 - 3\ 000}{0,40 - 0,20} = \frac{2\ 000}{0,20} = 10\ 000 \text{ tonne}$$

Table 2.7 Example on volume of transport for which the aggregated cost of the road and of the transport breaks even.

Road Class	Road Cost $/km	Speed of travel km/h	Variable transport cost $/t · km	Break even volume tonne
I	3 000	8	0,40	
				10 000
II	5 000	16	0,20	
				20 000
III	7 000	32	0,10	
				60 000
IV	10 000	64	0,05	

To adapt road standards to the volume of traffic is as we have seen in principle easy. However, in practice this problem is much more complex as we have to consider all traffic use of the road and also to decide the time horizon and the realistic traffic volumes for the chosen time period. In the next chapter we shall analyse in detail real life application of the formula (2.17).

The graph figure 2.7 illustrates this road standard exercise.

2.6.3 Modes of Transport at Varying Terminal Costs. The cost of loading and unloading is often specific for each mode of transport. Usually high terminal cost is combined with low variable transport cost and vice versa. At a certain transport distance the cost curves break-even. The break-even point can be mathematically derived provided that the two cost curves can be established as shown in the following example for two modes of transport, 1 and 2.

$$TC_1 = CT_1 + vdc_1 \cdot S \qquad (2.18)$$

$$TC_2 = CT_2 + vdc_2 \cdot S \qquad (2.19)$$

$$S = \frac{CT_1 - CT_2}{vdc_2 - vdc_1} \qquad (2.20)$$

TC = total transport cost
CT = terminal cost
vdc = variable transport cost
S = distance of transport

In this example it is assumed that no road costs are involved, but there will be no difficulty to introduce such items.

This introductory analysis of transport systems reflects the rule that in practically all work processes there are fixed and variable costs involved. The terms fixed and variable denotes that in each work operation a fraction of the cost is independent of the size of the operation expressed in number of units produced. The other share of the total cost is directly correlated to the numbers.

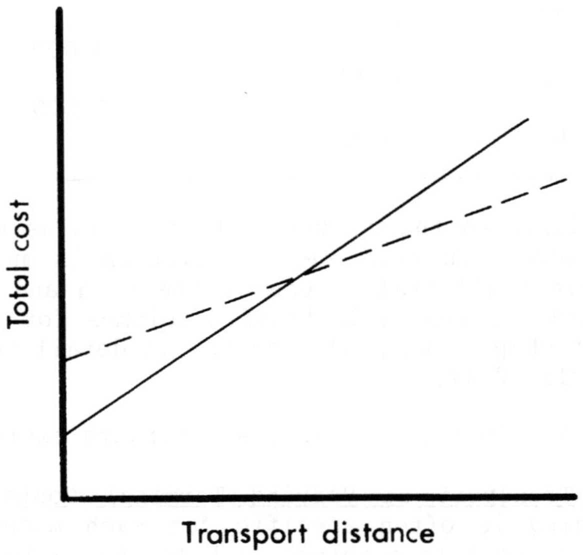

Figure 2.8. Break-even point for two modes of transport with high/low terminal cost resp. low/high variable transport cost.

2.7 LINES SERVING AREAS - A ROAD NETWORK

Forest roads are indispensable for the use of the resource. They penetrate the forest area and form a rather even net work, the pattern of which is influenced by a number of factors such as topography and flow of timber transport. The cost of forest roads is substantial. The investment for their construction may often amount to hundred dollars per hectare or more and the cost of maintenance per unit area is often of the same magnitude as the cost of silviculture. It goes without saying that the planning, design and maintenance of the forest roads is one of the most important managerial tasks.

The benefits of forest roads are manifold. Not only is timber transport facilitated, but also labour and supervisory personnel use them, the control of forest fires and pests can be more effective and hunters, berry pickers and nature lovers travel on them. - We shall now regard them as a system of lines serving the forest area in form of two different functions.

Timber, personnel and materials move on the road, which can be termed the lengthwise function of the roads. But roads also serve as the point of delivery for the preceeding timber transport or as points of departure for labour and material getting into the area, the crosswise function. Often the term drainage area is used to connote the area for which a road has a crosswise function, in analogy with rivers and waterways.

In most cases the roads with crosswise function are also points where a change from one mode of trnasport to another takes place.

This pattern is very significant for all transport systems involving collection or distribution to areas. People move from their homes to a network of feeder busses for a sub-way or rail line, mail is collected and distributed in a dual network with assorting depots organized along lines from which a crosswise function is exercised. It shall already now be observed that - once a road system meets certain basic requirements for its lengthwise function - the standards of the road has no influence on its crosswise function. An improvement of the road standard has no impact on the tecnology or cost of the linked transport. In contrast, the transports on the road, the lengthwise function, is directly influenced by the standards of the road. High costs for crosswise transport to a sparse road net of low standard can partly be offset the lengthwise transport on these roads of higher standards (because of higher traffic volume) but none of these functions are then optimally organized.

The design of road nets consists of two problems - optimum spacing of the road net and optimum standards of its different sections.

2.7.1 Optimum Spacing or Density of a Road Net. The more widely apart roads are spaced in an area the longer are the transports leading up to the road and the larger is the area that each section of a road serves. This is shown in figure 2.9 in which a road section and its tributary area is shown. The cost of the crosswise transport then is

$$C_{cd} = C_t + a \cdot \frac{S}{4} \qquad (2.21)$$

C_{cd} = cost of the crosswise transport per unit of product

C_t = terminal cost per unit of product

a = variable distance cost

S = the spacing to next roads

The area served by the section per unit length of the road is $S \cdot 1$ and, if Q product units derive from each unit area the total quantity is $Q \cdot S$. Setting the cost of the road per unit length to C_R, the cost per unit product, which we classify as the indirect transport cost (C_{ci}), is

$$C_{ci} = \frac{C_R}{Q \cdot S} \tag{2.22}$$

The total cost then is

$$C_{cd} + C_{ci} = C_t + a \cdot \frac{S}{4} + \frac{C_R}{Q \cdot S} \tag{2.23}$$

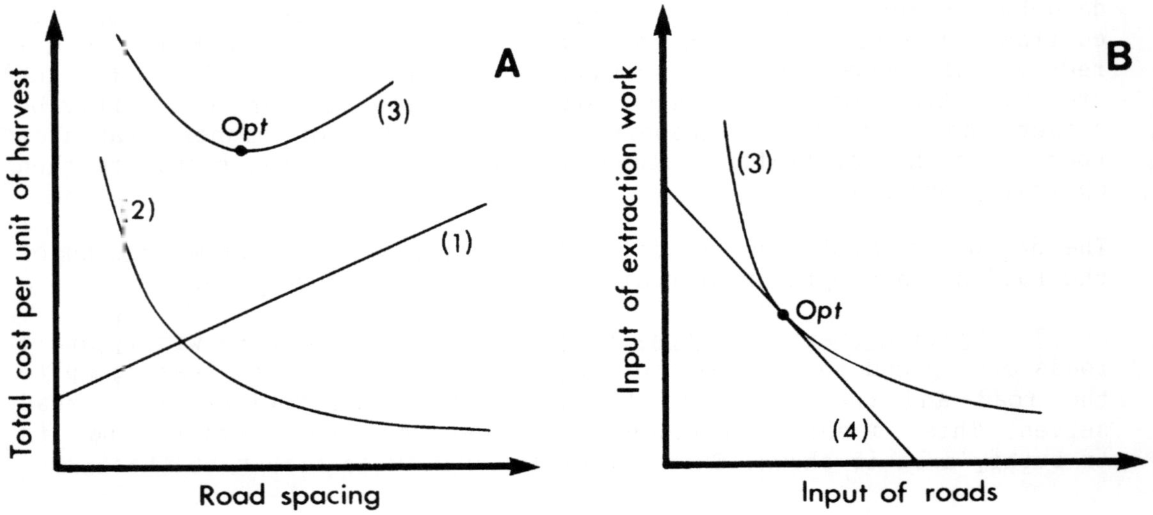

Figure 2.9. Optimal road spacing, according to the model in formula (2.23) in graph A, and using the isoquant in formula (2.34) in B.

Graph A

(1) Cost curve for extraction to roadside

(2) Cost curve for roads

(3) Aggregate cost of (1) and (2)

Opt = Optimum spacing

Graph B

(3) Isoquant

(4) Cost line

Opt = The physical inputs of roads and extraction work, resulting in minimum aggregate cost

The first differential of this function is

$$f' = \frac{a}{4} - \frac{C_R}{Q \cdot S^2} \qquad (2.24)$$

Setting f' = 0, to find the minimum cost and solving S, we get

$$S = 2 \sqrt{\frac{C_R}{a \cdot Q}} \qquad (2.25)$$

which is the mathematical expression for optimum spacing.

Another method of deriving optimum road spacing is to use the isoquant model, commonly used in engineering economics

$$\bar{q} = f(v_1, v_2) \qquad (2.26)$$

v_1 = input of roads per unit product
v_2 = input of extraction work, stump to roadside

The isoquant has a hyperbolic form as the rule of diminishing returns for increasing inputs of v_1 and v_2 applies. The curve measures the ratio at which the factors can be substituted against each other.

The cost line is

$$TC = p_1 \cdot v_1 + p_2 \cdot v_2 \qquad (2.27)$$

TC = total cost of producing one unit
p_1 = price of one unit of the input v_1
p_2 = price of one unit of the input v_2

We transform this function to

$$v_1 = \frac{TC}{p_1} - \frac{p_1}{p_2} \cdot v_2 \qquad (2.28)$$

The slope of the cost line is

$$\frac{dv_1}{dv_2} = - \frac{p_2}{p_1} \qquad (2.29)$$

TC must also satisfy the isoquant. At the tangential point one gets the input mix for minimum costs. The differential of the isoquant (2.26) is

$$d_q = f_1 dv_1 + f_2 dv_2 \tag{2.30}$$

$$\frac{dv_1}{dv_1} = - \frac{f_2}{f_1} \tag{2.31}$$

It follows that

$$\frac{f_1}{f_2} = - \frac{F_1}{F_2} \tag{2.32}$$

Road input $(f_1) = \frac{1}{S \cdot Q}$ per unit of Q

S = spacing between the roads
Q = extracted quantity per unit area

Extraction input $(f_2) = \frac{S}{4} \cdot a$ per unit of Q

Assume: road cost per unit length of road = C_R

extraction cost (variable component) per unit length and unit of Q = a

The formula (2.33) will then read

$$\frac{\frac{S}{4}}{\frac{1}{S \cdot Q}} = \frac{C_R}{a} \tag{2.34}$$

$$S_{opt} = 2 \sqrt{\frac{C_R}{a \cdot Q}} \tag{2.35}$$

which is identical with the previously derived formula (2.25).

The isoquant approach is a commonly used model in economics but has been little used by researchers in forestry. It is presented here as the basic model offers good opportunities to extensious into more complex problems as studies of publications in economics will reveal to the reader.

The isoquant approach for optimum road spacing has been shown because it is commonly used in engineering economics for studies on the micro level of production processes and can be used for many forestry problems. Another application will be made in Chapter 4 dealing with the optimum level of mechanization.

We now have a simple formula through which the optimum spacing of a road net can be derived. However, as the model is greatly simplified a number of practical problems come up when applying it in real life situations. These will be further looked into in Chapter 3.

2.7.2 Road Standard. The same problem structure is encountered for the determination of optimum road standard. A road of low standard will allow only low speed and high cost of moving the timber where as a high standard will give high road cost but low cost of the proper transport. The objective is to find a road standard which minimizes the aggregate cost of these two cost component.

Assume a straight line correlation between road cost (RC) and speed of travel

$$RC_i = a + b \cdot v_i \tag{2.35}$$

RC_i = Road cost for road standard i

v_i = average speed of vehicle on road standard i, for roundtrip

a and b = technical constants

This straight line relationship seems to be reasonable realistic for speeds up to 70 km/h.

Assume further that the cost per unit time of the vehicle, incl. operator is lc_i and independent of speed of travel (v_i), also a reasonably realistic assumption, the cost of moving (MC_i) the total product quantity (Q)

$$MC_i = \frac{lc_i \cdot Q}{q_i \cdot \frac{v_i}{2}} \tag{2.36}$$

q_i = payload of the vehicle

$$RC_i + MC_i = C_i = a + b \cdot v_i + \frac{lc_i \cdot Q}{q_i \cdot \frac{v_i}{2}} \qquad (2.37)$$

The differential of C_i is

$$\frac{dc}{dv} = b - \frac{2 \cdot lc_i \cdot Q}{q_i \cdot v^2} \qquad (2.38)$$

To find the optimum numeraire of v we put

$\frac{dc}{dv} = 0$ and solve

$$v_{opt} = \sqrt{\frac{2 \cdot lc_i \cdot Q}{q_i}} \qquad (2.39)$$

As both lc_i and q_i can be assumed to be independent of v the formula can be simplified to

$$v_{opt} = \text{constant} \cdot \sqrt{Q} \qquad (2.40)$$

which means that the optimum road standard, quantified as average possible speed of the vehicle is proportional to the square root of the total transported quantity. In Chapter 3 the assumed straight line relationships in this model shall be further discussed and a number of problems in quantifying the parameters in formula (2.37) shall be elucidated.

Again we meet the magic square root. It derives from the hyperbolic relationship $\frac{1}{X}$, which is a mathematical expression of the rule of diminishing returns for increasing inputs. However, this rule does not always follow $\frac{1}{X}$ where X has the exponent -1. The hyperbola can have a more complex mathematical form but this does not change the basic structure of the problems. It only calls for more complex mathematics with which we shall not burden us here. What is assumed when postulating that the rule of diminishing returns applies, is that the first dollar for improving the road above minimum standard gives higher cost savings in the transport on the road than subsequent improvements.

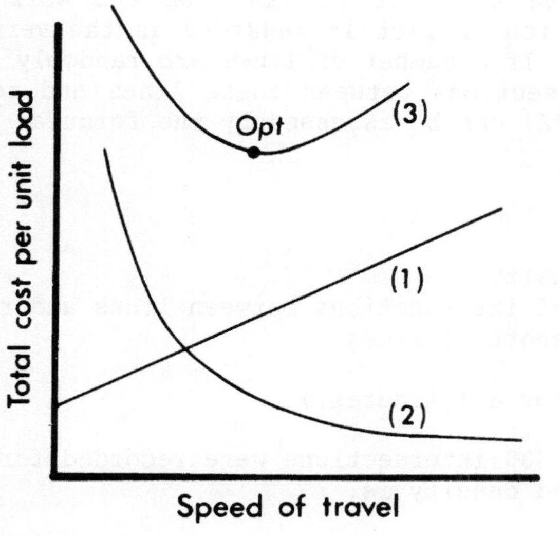

Figure 2.10. Optimal road standard, according to formula (2.37)
(1) = Road cost
(2) = Transport cost
(3) = Aggregate cost of 1 and 2
Opt = Optimal road standard

2.8 MEASURING ROAD DENSITIES

Roads are of paramount importance in forestry. Both the access roads leading up to the forests as well as the roads inside the forest boundaries - the feeder roads - are equally important. Data on the standards of all roads shall be collected. For the roads inside the forests also their total length in relation to the area - the density of the road net - is useful, as this ratio determines the average distance of transport from stump to roadside.

However, not all roads inside the forest boundaries have a crosswise function and some areas are unproductive. Such road sections and areas should therefore be recorded separately so that the road density is formed by the ratio between "road sections with crosswise function" and productive forest area. It is now standard practise to express road density in meter per hectare even if e.g. km per km^2 or miles per sq. miles is a more correct measure.

The road density of a forest area can be assessed in various ways. If good maps with the roads are available one can simply measure their total length on the map and divide this sum by the actual area.

If maps are not available the same measurements can be made on the ground. However, this is often a rather time consuming work and therefore a sampling method worked out by Matern and employed (von Segebaden, 1969) can be used. It is based on the work of Buffon - his needle problem - which in fact is regarded as the very first treaty of probability theory. If a number of lines are randomly laid out over the area and all intersections between these lines and roads are recorded the road density (RD) can be assessed by the formula

$$RD = \frac{\pi \cdot n}{2 \cdot L} \qquad (2.41)$$

where RD = road density
n = number of intersections between lines and roads
L = total length of lines

(Note: Same units for all figures.)

Assume a case where 80 intersections were recorded for a line length of 50 km. Then the road density is:

$$\frac{\pi \cdot 80}{2 \cdot 50} = 2{,}5 \text{ km/km}^2 \text{ or 25 meter per hectare}$$

The lines shall be randomly laid out but if the directions of the roads is randomlike and not of a parallel pattern, there should be little bias introduced in using an uniform line system.

This method is used in many research contexts both in laboratories and in the field. The properties of fibres can be determined, the density of a game population and similar assessments can be made, provided certain stochastic conditions are met. von Segebaden assessed in the Swedish National Forest Survey the whole forest road net in Sweden by letting the inventory crews record the intersections between their cruising lines and roads. By taking note of the road properties at each intersection a break down on road standards were also obtained. The results of such studies on existing road densities are very useful for quantifying the accessibility of the forests and also for assessments of the need for extending or improving the road net.

The method can also be used for the study of densities of skidding roads in logged areas and the soil disturbance made. It can also be used for assessing the amount of logging slash on a logged area. In the latter case, the length properties of each intersected slash component need to be recorded.

2.9 ACCESSIBILITY OF A FOREST RESOURCE. SITUATION FACTORS DETERMINING TECHNIQUES AND COSTS

2.9.1 Inventories and Data Needs. Any use of a forest resource is based on some survey of its wealth. A forest inventory, using some sampling technique is the standard approach for assessment of this wealth and its potentials. The inventory also is the data base for management planning.

By tradition, only parameters, significant and useful for silviculture and marketing are measured and recorded in forest inventories, such as tree species, tree size, quality, site conditions from the point of soil productivity etc. Only to a limited extent are they useful for assessment of accessibility, logging planning and for the costing of the various jobs. Such assessments are now growing more and more important requiring a broader data base with additional factual information, necessary for the technical planning and costing.

Such plans have different purposes, requiring various degrees of precision and refinement. It is common to distinguish between strategic, tactical and operational planning which recognizes the length of the planning period and the methods of assessing the inputs of manpower, machines and capital for the prospective use of the forest resource.

The ideal procedure is to include in a forest inventory of the usual type also the sampling of data of importance for the technical plans. The main components in such plans are road construction, logging and transport but it may also include manpower requirement and the technical planning of a set of silvicultural jobs. Inventories including also technical parameters are made - although often incomplete - in many countries, in particular in the national forest surveys in the Nordic countries. Many private forest firms nowadays have adopted inventory methods including the sampling of technical parameters.

However, it should be emphasized that there are considerable difficulties to work out an inventory method at reasonable cost, which meets the requirements both of the management planner and the logger. These difficulties originate in two sets of problems:

(i) The inclusion of technical parameters in an inventory often calls for a break down of the resource into strata which are technically consistent and meaningful. The number of strata of the inventory will increase requiring a much more dense sampling for a set level of precision. This problem is particularly apparent when technical data are needed for the short term (1-5 years) operational planning which requires rather detailed information for rather small areas.

(ii) The measurement of technical parameters may often be cumbersome and time consuming for the inventory crew, adding considerably to the cost of the inventory.

Very often, one is faced with a situation that inventory data of the traditional type already are available. The field work is already completed and a renewed field sampling is too expensive. Complimentary data for the technical planning and costing must then be collected from other sources such as maps, geological and metrological surveys, private firms or any other useful source of information. Such a situation is not unusual in project planning in developing countries.

As no comprehensive methodology exists for an appraisal of parameters relevant for working techniques this chapter will simply review some approaches which might be useful to employ. The success of an technical appraisal or costing exercise will ultimately depend on the skill of appraiser in using the set of data available in each specific situation.

2.9.2 Accessibility. A forest resource is often stratified as "Accessible Forests" and "Inaccessible Forests". The term accessibility is then as a rule used in an economic sense, although this is not always clearly spelt out. Given enough time and disregarding cost no forest is inaccessible.

Accessibility then connotes that there is or will develop a stumpage value on the wood. However, recognizing the fluctuations on market prices on wood and wood products it is evident that the border line between what is accessible or not from the point of view of economics will be moving back and forth as prices fluctuate. In Finland one uses the term Zero-Zone marking the extreme locations of this border lines at high and low market prices.

Accessibility can also be expressed in physical terms in units such as distance to nearest truck road and distance to a mill or export harbour. These distances can then be further broken down with regard to cost determining qualities such as topography, road standard etc. Also a stratification of the forest resource with regard to location to populated areas, villages and towns is often also meaningful as it will influence possibilities of recruiting and catering for the labour force and its dependents.

It seems advisable to use accessibility as a generic term for physical or economic location with regard to market place for forest products or to population centras. Inaccessibility is only meaningful at a certain time at specified price and cost levels.

2.9.3 Measuring location. As already pointed out the location of a forest resource is measured in the following distances:

(i) from stump to upper landing at existing or planned forest roads

(ii) from upper landing to processing site, broken down as feasible on private and public roads, rail or river or at any convenient intermediate point of significance for the transport operations.

These distances can be further subdivided in strata as terrain types, road standards or any other quality class influencing cost.

When inventories are made for very remote or non-developed areas it may be useful to specify the location both to existing facilities (i)-(ii) and to a planned development of main transport routes, mills etc. The principle is that if or when the resource is being developed a minimum of transport routes are absolutely needed for opening up the area and for providing a continuous access to the area. This main road system, which can be drawn up on a map before the inventory starts could consist of two parts. An access road from the outside to the perimeter of the forests and then main artery roads branching out in the forest area. It is suggested that the main artery roads should have a density of 3-5 m per hectare. When the forest then gradually is being used the artery road system is build out with secondary and feeder roads up to a density which in many cases will be 15-25 meter per hectare (see further Chapter 3.5). In the cost computation the access roads and the main artery roads can be looked upon as fixed overheads in using the resource whereas the secondary and feeder road system will be a variable cost item for the actual operations, mainly logging and transport.

In inventories it is desirable that these location factors are recorded for each sample unit. The reason is that the location often has influenced the actual stocking or state of the forests. Forest close to existing roads may often already be logged more heavily than the remoter parts or they may be better managed, grow on better sites or be subject to any biased influence. As the purpose of the location analysis is to give a true picture of the potentials of the resource such possible sources of bias must be carefully considered before the onset of the inventory and be catered for in the inventory design.

Once these location parameters are recorded there are various possibilities to compute them. Often it is meaningful in costing to delineate the forest property in sub-units, preferably drainage areas, of a size which is meaningful for the planning of forest road nets or logging operations. The sample units within each such sub-area then represents the whole sub-unit. However, sampling densities in inventories are often so low and the samples in each sub-area are so few that the standard deviations on stand and tree parameters will become unacceptable. There are various statistical designs which can alleviate to a considerable extent such short-comings. The whole forests may be stratified in type stands and the gross averages for tree and stand parameters for each such type stand can then be used for the single sub-units. The point is the importance of close counselling between the inventory experts and the logger or engineer in designing an inventory. The problem of appraising the technical parameters and the use of them for costing purposes should be clearly explained for the statistician, who then will be in a position to work out an inventory design meeting as far as possible the data need of the technical planner.

An outline of the main steps for accessibility studies in conjunction is presented in table 2.8 for two situations. The schemes should be looked upon as examples only as each forest is specific and each inventory also has specific objectives. However, the schemes reflect the principle.

OUTLINE OF PROCEDURES FOR INTEGRATING THE ACCESSIBILITY IN FOREST INVENTORIES

PHASE	Alternative I Forest with established transport routes	Alternative II Forest without road net, often remote
PLANNING AND DESIGN	Locate on maps roads, towns, mills and other points for delivery of products.	Outline on maps access routes to the perimeter of the forest and main roads in the area to a density of 3-5 meters per hectare.
	Design inventory and include methods and procedures for measuring and recording physical logging and accessibility variables.	
FIELD WORK CRUISING	Record for each sample unit physical variable for trees, stand, terrain and soil relevant for logging and road construction	
	Record for each sample unit accessibility, e.g. distance to nearest road and point of delivery of forest products and to towns.	Assign identification of each sample unit for its location on maps in next phase "Office work".
OFFICE WORK	Develop cost models for logging and transport and for road construction and maintenance. Assess optimum road densities for actual exploitation and management schemes and break down roads net on road standards.	
	Assess cost for logging, transport, roads etc for two systems (i) using only existing roads (ii) for operating with optimum road density.	Locate sample units on maps and measure straight line distances to main roads as drawn up on maps and to points of deliveries. Apply T factor for computing to real distances.
COMPUTATION AND PRESENTATION	Stratify the forest resource in biological, utilization, accessibility, sub areas or other strata as desired and present results accordingly. Compare for each strata as desired and present results accordingly. Compare for each strata unit cost for delivered products with unit prices. If applicable, establish cost limits or "zero stumpage" lines and locate them on maps and draw up accessible and inaccessible forests areas.	

3. APPLICATIONS AND QUANTIFICATIONS

3.1 INTRODUCTION

In the previous chapter conceptual models of some main work operations were worked out and the main factors or variables bearing on the production rate or cost identified. The conceptual models were formalized in mathematical equations with the purpose of generalizing the problem and of allowing abstract study of reality under a wide range of work conditions. The ultimate goal is to find optimal solutions with regard to input/output relationships. The mathematics is by purpose as simple as possible which will enable the student to focus his attention on the problem per se rather than be distracted by complex mathematics.

The next phase is the collection and computation of data with relevance and validity for the study. This is often a much more difficult, tricky and time-consuming task than the modelling of the problem. The widely varying work conditions in forestry tend to make data from different sources incomparable and often also inconsistent. It often requires both skill, experience and good judgement to provide the quantifications of the parameters in the abstract models.

It is our experience that the quantification of inputs and outputs are of greater importance than to refine the models with scores of variables which may or may not bear on the problem. It is definitely preferable to retain simplicity of the models then to aim for perfect fitness between a set of observation and a complex regression function. However, in the Appendix B to Chapter 3 some such functions are exemplified.

A word of caution needs to be said. In reality, the job problem may often be more complex and influenced by other variables than by those used in these cases. But as our purpose is purely pedagogic only such variables will be included which determine the general form and the structure of the case. When using such mathematical models for practical purposes more care may be needed and more advanced mathematics may be required so that the model reflects more fully the job relationships. This is only a formal extension of the model. Once the derivation of the simple model is made and understood it will not be difficult to adjust it more closely to reality or to enlarge it by inclusion of more variables or more complex relationships between the independent and the dependent ones.

3.2 MODELS OF LINEAR TRANSPORT

3.2.1 <u>Transport Work and Travelled Distance.</u> Moving a quantity, q, a distance of s results in the transport work of q · s. In transporting forest products the vehicle often travels empty on the return. For each roundtrip the vehicle then travels the distance 2 · s. In the extraction of logs from the forest to roadside the tractor often in one roundtrip picks up logs at various points. Then the transport distance of the cargo will differ from the distance travelled by the vehicle. It is therefore sometimes useful to distinguish between transport distance and roundtrip distance.

Assume that a forwarder travels empty 400 m and turns. It then loads 4 tonne each at 400, 360 and 300 m distance from the road. Then the transport work is:

$$
\begin{aligned}
0{,}4 \cdot 4 &= 1{,}6 \quad t \cdot km \\
0{,}36 \cdot 4 &= 1{,}44 \quad \text{"-} \\
0{,}3 \cdot 4 &= \underline{1{,}2 \quad \text{"-}\qquad} \\
&\ 4{,}24 \ t \cdot km
\end{aligned}
$$

The average transport distance of the logs: 4,24 : 12 = 0,353 km
The roundtrip distance of the forwarder: 0,4 km

Thus, transport distances of the cargo is not always the same as roundtrip distance.

This may have practical implications in some cases. Assume a logging area where two assortments are taken out. The operator want to have the assortments separated at roadside. He then has the following five options:

A Assortments piled separately by feller		B Assortments mixed in piles.		
A_1	A_2	B_1	B_2	B_3
Forwarder hauls each assortment separately.	Forwarder hauls both assortments on same roundtrip but separated on the load platform.	Haul mixed. Separate in unloading.	Separate logs in loading, assortment hauled together.	Load and haul separately.

Assume 24 tonne of logs of two assortments, each 12 tonne in three piles of 8 tonne on the distances 400, 360 and 300 meters to be hauled by a forwarder with 12 tonne load capacity. In the cases A_1 and B_3 the roundtrip distance will be 400 meter, whereas in the cases A_2, B_1 and B_2 it will be 350 m. The excess distance of 50 m in cases A_1 and B_3 is the cost of having logs separated for these two methods. In practice, an operator does not have to adhere to one of the methods but use them as he finds best in each situation.

The conclusions are:

(i) Average transport distance and average roundtrip distance is only equal when the vehicle is fully loaded at one point.

(ii) Loading the vehicle at different points the following approximate correlation can be used:

Av. roundtrip distance = Av. transport distance + av. distance travelled in loading

(iii) To keep the roundtrip distance as short as possible the operator shall try to get a full load as quickly as possible.

(iv) Taking out several assortments in separate roundtrip will lengthen the average roundtrip distance, except in the hypothetical case, when a full load can be achieved of each assortment at one loading point only.

3.2.2 Transport Along Lines. The cost of transport is related to the distance. Most transport enterprises have tariffs or tables with the cost per unit cargo for each transport distance. Collective wage agreements often include similar tariffs. Such tariffs are the results of experience of the entrepreneur or by an analysis and detailed costing of the operation, or of both.

Studying such tariffs, one often finds that the cost increases linearly with the distance. The tariff can then be transformed in the mathematical formula of the straight line:

$C = a + b \cdot s$
C = cost (3.1)
S = distance of transport
a and b = coefficients

As it is useful for planning purposes to have the relation between transport cost and distance in a mathematical formula, some exercises follow on the derivation of such formulae.

In three different regions the following tariffs for truck hauling of timber are used:

Table 3.1 Example of tariffs of transport

Distance, km		20	30	40	50	60	70
Cost of transport, US$/m^3,	Reg I	3,00	3,30	3,60	3,90	4,20	4,50
	Reg II	3,00	3,30	3,65	3,90	4,15	4,50
	Reg III	2,80	3,20	3,60	3,95	4,25	4,50

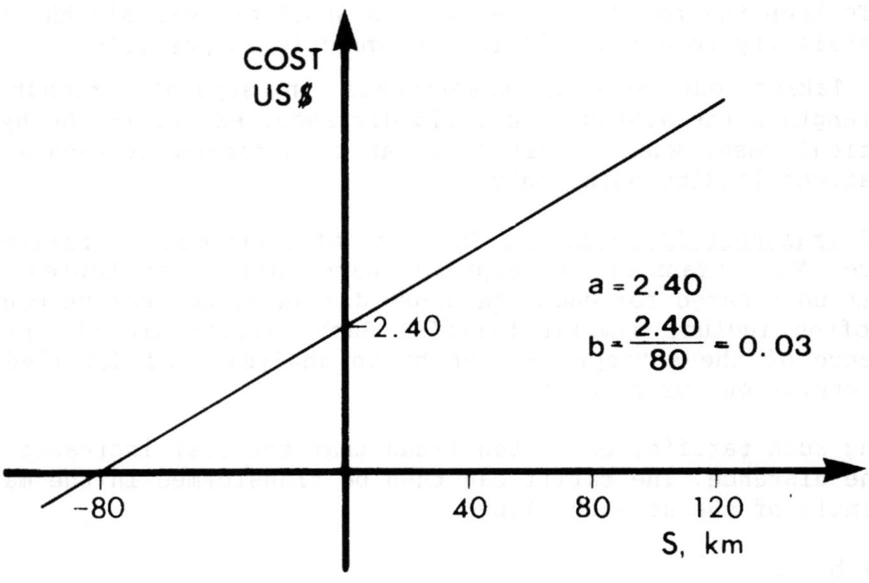

Figure 3.1. Determining "a" and "b" through plotting observations in a graph (Costs for Region I.)

Studying these cost series we find that for Region I the cost increases with exactly 30 cents for each 10 km, in other words a straight line relationship. There are several ways of deriving the mathematical formula. One can simply plot the figures on a plotting paper and draw aa straight line through the points as shown in figure 3.1. By reading the intersections between this line and the coordinate axles the cost curve is established:

$$C = \frac{2,40}{80} \cdot S + 2,40 \quad \text{or } C = 0,03 \cdot S + 2,40$$

where S is the distance in kilometer and C the cost in dollars per m^3. The formula tells us that for each one kilometer increase of distance there will be an cost increase of 3 cents.

The formula is valid for distances between 20 and 70 kilometers. Extrapolations should be made with care.

Looking on the tariff for region II we find an irregular cost increase for each 10 km class, in cents: 30, 35, 25, 25, 35. If we find no logical reason for these irregularities, we assume that the tariff is the result of a careless computation or is some kind of a compromise in the negotiation between truck haulers and firms. The same cost line as for region I represents the relationship fairly well and we may assume that it better reflects the true cost than the original tariff.

The tariff for region III shows decreasing cost increment for each 10 km class: 40, 40, 35, 30, 25. Thus a convex curve better represents the relationship than a straight line. We shall in this text refrain from using more complex mathematical expressions as they easily will obscure our purpose - to explore and develop theories and principles. Such simplifications are a formal deviation from reality. It does not conflict with the basic approach of using mathematical models for analytic purposes but it infers a certain discrepancy in our study. In practise, one has to choose between accepting a certain discrepancy in keeping the mathematics on a managable level or to make the analysis so mathematically complex that only computers and professional mathematicians can deal with it. In the case of region III it may for example be so that we are only interested in cost differentials in the range of 50 to 70 kilometers. Instead of introducing a complex formula we can in such a case proceed in two ways:

(i) using the average of the numeraires between 50 and 70 km, 30 and 25, in the straight line formula which then will read $C = 0{,}0275 \cdot S + a$, or

(ii) carrying out two computations using the formula: $C = 0{,}03 \cdot S + a$ and $C = 0{,}025 \cdot S + a$

The true answer will then be between the results of these two computations, which many times is a satisfactory answer for practical purposes.

The term "a" is often of less significance and its value not derived here.

Let us now return to the tariff for region I which was transformed to the formula $C = 0{,}03 \cdot S + 2{,}40$. The coefficient 0,03 is then the variable distance cost of the transport as we defined it in Chapter 2. Similarily, 2,40, is the terminal cost of the transport. It may include not only the actual cost of loading and unloading but also the time cost of turning the vehicle, get it ready and similar terminal work related to the transport and also the time consumed in accelerating and breaking (retarding) the vehicle on departure and arrival in excess of time at normal speed.

A formula for transport cost can also be derived from time studies or time records. The variable distance cost obviously is:

$$\text{Variable distance cost} = \frac{\text{cost of vehicle per unit time}}{\text{payload} \cdot \text{av. speed of transport}}$$

A truck loading 30 tonnes of logs travelling with an average speed of 50 km on the roundtrip (return empty) performs a transport of

$$\frac{30 \cdot 50}{2} = 750 \text{ t} \cdot \text{km/h}$$

We make an estimate of the cost per hour for the truck and the driver in travelling and arrive at a cost of 75 US$ per hour. Then the variable distance cost is

$$\frac{75}{750} = 0{,}10 \text{ US\$ per t} \cdot \text{km}$$

3.2.3 <u>Cost of Commuting.</u> Transport of labour incurs cost. Sometimes the labour is required to travel outside the shifthours, sometimes the travelling takes place within shifthour. Whatever the compensation, the travel costs money. Even when travels are made outside the shifthour it represents a cost for the traveller as he always puts some value on his free time. An employer often must compensate for that in one way or the other. In principle the travelling cost of persons will depend on the value per unit time that is accorded to the person or that he himself puts on his time plus the cost of vehicle. The same reasoning applies not only to labour but to supervisors and managers as well. As forest operations require a lot of movements of personnel the cost of travel should be kept as low as possible, both for the enterprise and for the traveller. In fact, the cost of moving personnel is so high that it needs to be carefully considered in the design of forest road nets. The following example illustrates the case.

Assume a forest area of 1 000 hectares with an annual yield of 5 m^3 per hectare. The labour productivity is 10 m^3 per day for all work in the area. The workers commute daily. We now put two time values for the workers. If the travel is made on shift time it will simply be the cost per hour for labour, say 8 US$ per hour in direct pay and 4 US$ in fringe benefits, in total 12 US$ per hour. Travelling on free time we use a time value of 75% of 12 US$ = 9 US$ which is a common practice in traffic studies. The speed of the vehicle is 60 km/h. We now make the following computation.

	Time value per hour	
	US$ 12	US$ 9
Travel time per km of road	2 minutes	2 minutes
Value of travel time	$\frac{2}{60} \cdot 12$	$\frac{2}{60} \cdot 9$
Production per worker day	10 m^3 = 8 tonne	10 m^3 = 8 tonne
Unit cost of travel	$\frac{2 \cdot 12}{60 \cdot 8} =$	$\frac{2 \cdot 9}{60 \cdot 8} =$
	= 0,05 US$ per t · km	= 0,0375 US$ per t · km

The cost of labour transport is in this case 5 (or 3,75) cents per t · km, which is close to a normal cost of truck hauling of logs, in spite of the fact that only the time value of the worker is included. If we add the cost of the vehicle - a bus or a car - the cost of transport of labour will in many cases be higher than for transport of timber.

Many forest jobs are carried out in the forest area requiring the worker to walk from the road to the work place. Examples are felling, planting, cleaning, pruning etc. The walk is made once or several times per shift depending on how the provision of food, drinking water and other needs is catered for. Assume that the harvesting of 100 m^3 (= 80 tonne) requires the input of 8 such area bound mandays (10 tonne per manday) and that the workers walk with a speed of 3 km per hour. The time of a roundtrip distance of 1 km is

$$\frac{2}{3} = 0,67 \text{ hour}$$

Using a time value of 12 US$ per hour the cost is

$$0,67 \cdot 12 = 8 \text{ US\$ per km}$$

Now, each of these tonnes shall be extracted. It is therefore possible to express this cost in terms of transport work in order to be able to aggregate it with the variable cost of the actual log transport. This variable cost for the movements of the workers from the road is then, in terms of transport work

$$\frac{8}{10} = 0,80 \text{ US\$ per t} \cdot \text{km}$$

as the worker produces 10 tonne per shift.

(i) The cost of travel on roads and walking from roadside to workplace in the forest can be assessed and expressed in the same units as for timber transport, e.g. in \$ per t · km. The purpose of this is to aggregate the costs of all movements on a road or in a forest area, which is a necessity in the optimization of forest road nets.

(ii) The actual cost will depend on the time value of labour in commuting. If the moving takes place on shift time, this time value should be the same as or close to the actual cost per unit time for labour. If the moving is made on free time - outside the shift - it is reasonable to use a lower time value. In studies on the economics of public highways it is common to use a time value of around 75% of the normal cost per hour for ordinary labour. This is always an arbitrary figure which can or should be considered with regard to the actual situation on the labour market. If labour is plentiful with low marginal cost for hiring additional workers, there are good reasons to use a low time value, whereas a higher time value can be advocated in situations of labour shortage with high marginal cost of increasing the labour force. The time value used may also depend on the purpose of the study, e.g. if its made for finding the best economics of transport for a private enterprise or for public transportation.

(iii) Comparing the actual cost of labour transport in these assessment with moving timber, it is evident that it is considerable. In fact, the variable distance cost for labour for the movement from roadside to the forest area is often in the range of one third of that of moving timber. For movements on a truckroad the transport of labour often costs more than the actual transport of timber. The practical conclusion is that transport of labour must be considered in the planning of forest roadnets both with regard tc the spacing of the roads and the choice of road standards.

3.3 BREAK EVEN POINTS FOR MODES OF TRANSPORT

Timber flows with few exceptions in one way only. The vehicle travels empty on the return. Often speed of travel is different loaded and empty. Observe that the average speed of a roundtrip (vΠ·≡) is

$$\frac{2}{v_{ave}} = \frac{1}{v_{loaded}} + \frac{1}{v_{empty}} \qquad (3.2)$$

As explained in a previous chapter the aggregate cost, C, of moving timber over a distance, S, can often be expressed in the formula for a straight line:

$$C = a + b \cdot S \qquad (3.3)$$

where "a" is the cost of loading and unloading - the terminal cost - and "b" is the variable distance cost. The parameters a and b are different for each mode of transport. The formula above can be used for a general survey on what range of distance a certain mode of transport is suitable. Assume that a sawmill have three options for the transport of timber on roads: farmtractor with trailer, a truck and a truck with trailer. The cost parameters of the three modes of transport are as follows:

	Terminal cost	Variable cost b
Farm tractor with trailer (FT)	1,90 US$/tonne	0,22 US$/t · km
Truck (T)	2,50 "-	0,08 "-
Truck with trailer (TT)	2,80 "-	0,05 "-

This cost relationships can be plotted in a graph, figure 3.2 and the break even points be read. The relationships can also be established in three formulas:

$$C_{FT} = 1,90 + 0,22 \cdot S \qquad (3.4)$$

$$C_{T} = 2,50 + 0,08 \cdot S \qquad (3.5)$$

$$C_{TT} = 2,80 + 0,05 \cdot S \qquad (3.6)$$

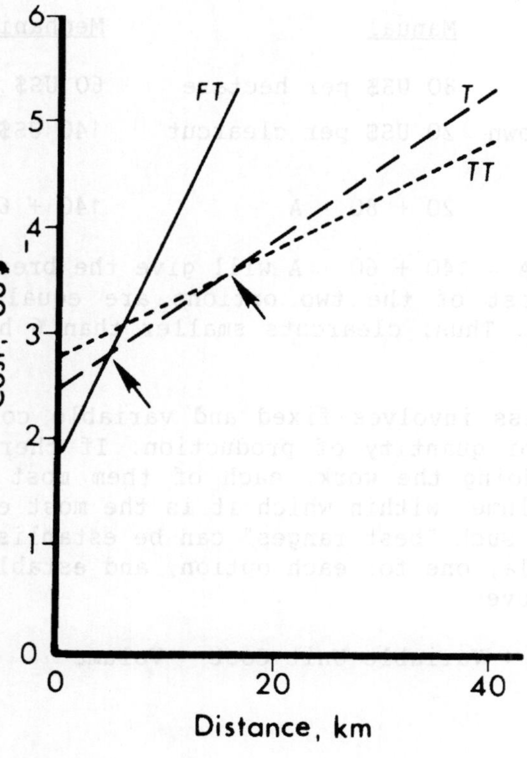

Figure 3.2. Break even points for different modes of transport.

The break even points are then; when $C_{FT} = C_T$ and $C_T = C_{TT}$

Farm tractor/truck $\dfrac{2,59 - 1,90}{0,22 - 0,07} = 4,3$ km

Truck/truck with trailer $\dfrac{2,80 - 2,50}{0,07 - 0,05} = 15$ km

The following rule of thumb then applies: Use a farm tractor with trailer on distances shorter than 4,3 km, a truck with trailer on distances over 15 km and a truck without trailer in the distance range of 4,3 km to 15 km.

Many forest jobs are bound to the forest area, such as planting, weeding, scarifying etc, implying that the total cost of the job is directly proportional to the total treated area. However, often there is a fixed cost for starting up and closing down the operation, which may include the transport of equipment to and from the area. Consider two options - manual and mechanized - for the scarifying of a clearcut area.

	Manual	Mechanized
Operating cost	80 US$ per hectare	60 US$ per hectare
Starting up & closing down	20 US$ per clearcut	140 US$ per clearcut
Total cost per clearcut of A hectare	$20 + 80 \cdot A$	$140 + 60 \cdot A$

The equality $20 + 80 \cdot A = 140 + 60 \cdot A$ will give the break even point, the A for which the cost of the two options are equal. Solving the equation A = 6 hectares. Thus, clearcuts smaller than 6 hectares shall be manually scarified.

Almost every work process involves fixed and variable cost components with regard to volume or quantity of production. If there are several technical options for doing the work, each of them most likely has a range with regard to volume, within which it is the most economical. As shown in these examples such "best ranges" can be established by applying the following formula, one for each option, and establish the break even points as shown above.

Total cost = Fixed Cost + Variable Unit Cost · Volume

or

$$\text{Unit Cost} = \frac{\text{Fixed Cost}}{\text{Volume of Production}} + \text{Variable Unit Cost} \qquad (3.7)$$

3.4 STATIONS SERVING LINES OR AREAS

In performing a forest job, a station or depot of some kind is often needed serving the surrounding area. Such a depot can be logging camp, an air-strip for fertilization from the air, a quarry, a central barking place or some similar installation. These cases have in common the question: How closely spaced shall these installations be?

To answer this question two analytic approaches are possible. One is to select three or more spacings - say with the distance between the depots of S, 2.S and .3S - carry out an estimate for each alternative and plot the results in a graph. The result will be a cost curve with a cost minimum for a certain value of S. The other approach is to make a mathematical model of the operation with the unit cost as dependent variable. By means of derivation of such a cost function the optimum solution giving minimum unit cost is found. A few examples will be elaborated.

Assume that a large forest tract shall be logged and logging camps are needed for the labour. It is planned that the workers shall walk once daily from the camp to the work place. How closely shall the camps be spaced. In the chapter 2 it has been shown that the camp in principle shall be located in the centre of a hexagon but that using a square as the area served by the camp does not make any great difference. To simplify mathematics we thus make a model of the whole tract in squares of equal size.The side of the tract, a, is then also the spacing between the camp sites. In the table 2.5 p. 64 we find that the average distance walking from the centre of a square to the area is 0,383 of the side of the square, a. The criteria for an optimum spacing of the camp is that the aggregate cost of moving the camp from one site to another and the cost of walking of the labour shall be as low as possible per unit of production. We then collect the following data:

Cost of moving the camp	(M)	1 300 US$
Productivity of the worker	(P)	12 m^3 per day
Density of cut	(Q)	180 m^3 per hectare (=18 000 m^3/km^2)
Time value of workers in walking, say 50% of an earning of 8 US$	(D)	4 US$ per hour
Speed of walk	(V)	3 km per hour

Assume a square with the side "a" km the following unit cost are derived:

Cost of moving the camp $\frac{1\,300}{a^2 \cdot 18\,000}$ \$ per m^3 or $\frac{M}{a^2 \cdot Q}$

Cost of workers' daily walk $\frac{0,383 \cdot a \cdot 2 \cdot 4}{12 \cdot 3}$ \$ per m^3 or

$$\frac{0,383 \cdot a \cdot 2 \cdot D}{P \cdot V}$$

The aggregate cost C then is:

$$C = \frac{M}{a^2 \cdot Q} + \frac{0,383 \cdot a \cdot 2 \cdot D}{P \cdot V} \qquad (3.8)$$

In deriving this formula we get

$a_{min} = \sqrt[3]{\frac{M \cdot P \cdot V}{0,383 \cdot Q \cdot D}}$ and using the parameters in the example

$$a_{min} = \sqrt[3]{\frac{1\,300 \cdot 12 \cdot 3}{0,383 \cdot 18\,000 \cdot 4}} = 1,2 \text{ km} \qquad (3.9)$$

In the optimum solution the aggregate C then becomes

$\frac{1\,300}{1,2^2 \cdot 180} + \frac{0,383 \cdot 1,2 \cdot 2 \cdot 4}{12 \cdot 3}$ or 0,05 + 0,10 = 0,15 US\$ per m^3

In this case the rule of thumb will be that the spacing of the camp shall be around 1,2 km. However, one is often somewhat uncertain with regard to the quantification of one or several parameters in such an estimate.Assume that in this case the time value of the workers is disputable. One can then make a sensitivity analysis using higher and lower time values, say 2 and 6 US$ per hour instead of 4 and arrive at the following result:

	Time value	Optimum spacing	Aggregate cost
Base case	4 US$ per hour	1,2 km	0,15 US$ per m^3
Low time value	2 "-	1,5 km	0,10 "-
High time value	6 "-	1,04 km	0,20 "-

The computation shows that the optimum spacing is not very sensitive to changes in time value. One may readily conclude that the spacing should be between 1,1 and 1,5 km. The same sensitivity analysis can then be made for the other parameters, e.g. what is the influence of changing cost of the moving of the camp or of density of cut. As the result is proportional to $\sqrt[3]{}$ parameter values (or the inverse) the influence of minor variations are minute.

A forest enterprise has a portable crusher processing gravel for the maintenance of the forest roads. The crusher is set up at suitable points along the roads and processes a quantity of gravel to last for the road maintenance for a predetermined period of time. The question is now: At what spacing shall these gravel depots be located to give the lowest possible cost of gravel spread out on the roads, assuming suitable material is readily available everywhere along the roads. What quantity of gravel shall be available at each depot?

To answer this question one must establish a simplified geometrical model of the road net. Let us consider five cases and assume the following:

Gravel requirement	150 m^3 per km of road
Variable trucking cost of gravel from depot to road	0,20 US$ per m^3 and km
Cost of moving the crusher	600 US$
Half the distance between depots	a km

On the basis of this information the table 3.2, p. 105 can be made.

Table 3.2 Estimate of optimal spacings of gravel depots

ANALYSIS					
Road sections meeting at the gravel depot	2	3	4	6	8
Pattern	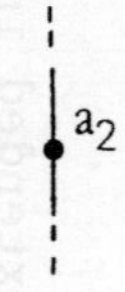	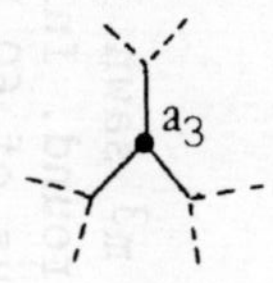	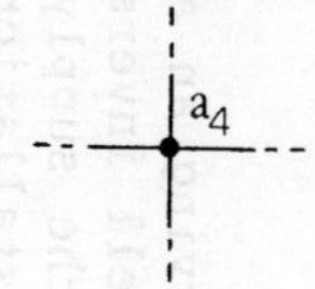	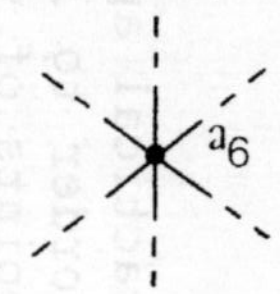	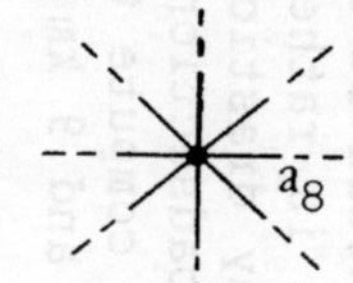
Av. transport distance of gravel	$\frac{a_2}{2}$	$\frac{a_3}{2}$	$\frac{a_4}{2}$	$\frac{a_6}{2}$	$\frac{a_8}{4}$
Road length served by each depot	$2a_2$	$3a_3$	$4a_4$	$6a_6$	$4a_8(1+\sqrt{2})$
Aggregate (partial) cost of gravel supply	$\frac{600}{2\cdot a_2\cdot 150}+0,2\cdot\frac{a_2}{2}$	$\frac{600}{3\cdot a_3\cdot 150}+0,2\cdot\frac{a_3}{2}$	$\frac{600}{4\cdot a_3\cdot 150}+0,2\cdot\frac{a_4}{2}$	$\frac{600}{6\cdot a_6\cdot 150}+0,2\cdot\frac{a_6}{2}$	$\frac{600}{4\cdot a_8(1+\sqrt{2})\cdot 150}+0,2\cdot\frac{a_8}{4}(1+\sqrt{2})$
a_n minimum	$\sqrt{\frac{600\cdot 2}{2\cdot 150\cdot 0,2}}=4,5$ km	$\sqrt{\frac{600\cdot 2}{3\cdot 150\cdot 0,2}}=3,7$ km	$\sqrt{\frac{600\cdot 2}{4\cdot 150\cdot 0,2}}=3,2$ km	$\sqrt{\frac{600\cdot 2}{6\cdot 150\cdot 0,2}}=2,6$ km	$\sqrt{\frac{600\cdot 4}{4\cdot 0,2(1+\sqrt{2})\cdot 150}}=$ $=1,6$ km
SOLUTION					
Spacing of depots:					
along the roads	$4,5\cdot 2 = 9$ km	$3,7\cdot 2 = 7,4$ km	$3,2\cdot 2 = 6,4$ km	$2,6\cdot 2 = 5,2$ km	$1,6\cdot 2 = 3,2$ km
straight line dist.	$4,5\cdot 2 = 9$ km	$3,7\cdot\sqrt{3} = 5,2$ km	$3,2\cdot 2 = 6,4$ km	$2,6\cdot 2 = 5,2$ km	$1,6\cdot 2 = 3,2$ km
Quantity of gravel per depot	$2\cdot 4,5\cdot 150=1350$ m^3	$3\cdot 3,7\cdot 150=1670$ m^3	$4\cdot 3,2\cdot 150=1920$ m^3	$6\cdot 2,6\cdot 150=2340$ m^3	$4\cdot 1,6(1+\sqrt{2})\cdot 150$ $= 2300$ m^3

Even if the estimate is a simplification of the forms of real road nets, it is rather safe to conclude that the distance along the road between the depots with the given cost parameters should be around 5,5 - 9 km. In a model like this the curve minimum cost is rather flat. A deviation of up to round ± 20% does not involve any drastic cost increases. Let us exemplify with the case of three roadsections meeting at the depot for which the optimum was 7,4 km. If we compute the aggregate (partial) cost of the gravel supply for 6, 7,4 and 9 km according to the cost formula we get the following result.

Spacing, km	6,0	7,4	9,0
Aggregate (partial) cost, US$ per m^3	0,74	0,71	0,75

These small cost differences indicate that in the practical application one can allow ± 20 on both sides of the optimum in order to locate the depots at points which are best suited from other points of view. Such sensitivity tests should always be done when transforming the result of an estimate to practical guidelines.

A forest mill can also be regarded as a station serving an area - the forester's view. The mill superintendent may very well inverse the relationship, the forest area servicing the mill by the supply of logs. The point is to balance the capacity of a fixed installation for processing a raw material drawn from a surrounding area of suitable size and form. This relationship shall be explored here only in its spatial context.

Assume a sawmill with an annual capacity of 20 000 m^3 sawn lumber requiring a log supply of 40 000 m^3 solid wood in the round. Initially we consider the supply area to be circular with a radius of 60 km, corresponding to an average transport distance (straight line or as the crow flies) of 40 km. It is planned to increase the production of the mill by 50% and an assessment shall be made on the implication on transport costs for the supply of logs.

(i) Assume initially that the supply area can be extended in all directions. Then the radius (R) of the enlarged supply area is derived from the expression:

$$\frac{R^2}{60} = \frac{40\ 000\ (1+\frac{50}{100})}{40\ 000} \qquad (3.10)$$

$$R = 60 \cdot \sqrt{1,5} = 73,4 \text{ km} \qquad (3.11)$$

The average transport distance will be $\frac{2}{3} \cdot 73,4 = 49$ km,

an increase of 9 km only (or 22,5%).

This finding can now be generalized for all configuration of the supply area provided the enlarged area is congruent with the initial one and there is homogenity with regard to yield per unit area - the same quantity is available per unit area on the new supply belt. Under such circumstances the following formula applies

$$S_2 = S_1 \sqrt{\frac{Q_2}{Q_1}} \qquad (3.12)$$

S_1 = transport distance for capacity Q_1

S_2 = "- Q_2

(ii) Let us now drop one of the postulates, equal yield per unit area and consider a case where only 25% of the yield per unit area in the enlarged belt can be expected due to competition from surrounding mills or for some other reason.

The supply can than be considered to consist of two circles one overlaying the other with the radia 60 km and R_{60+}. Then the supply from the larger circle will be

$$a \cdot \frac{25}{100} \cdot 40\ 000 = 10\ 000 \cdot a$$

log units/area unit and from the smaller circle, after deduction of the yield already supplied from the larger one

$$a \cdot \frac{75}{100} \cdot 40\ 000 = 30\ 000 \cdot a$$

log units/area unit ("a" is a constant factor)

Then, obviously both circles shall supply the same quantity and thus

$$\pi \cdot 60^2 \cdot 30\ 000 \cdot a = \pi \cdot R_{60+}{}^2 \cdot 10\ 000 \cdot a$$

and solving this equation we get

$$R_{60+} = 60 \cdot \sqrt{3} = 104 \text{ km}$$

The average transport distance is derived by weighting

$$\frac{30\ 000 \cdot 40 + 30\ 000 \cdot 40 \cdot \sqrt{3}}{60\ 000} = 55 \text{ km}$$

This is an example of modifying a model to correspond better to an actual situation. Several such modification can be made once the structure of the model is understood. - The same estimate can be made for supply areas in hexagonal form or any type of congruently extended areas.

(iii) In quantifying the cost increases, we first drop the "straight line" assumption and introduce a winding factor of 1,30. Variable truck hauling cost are assumed to be 0,05 US$ per km and m^3. The increase in mill capacity with 50% will then increase the truck hauling cost per m^3 of timber in the round as follows:

In case (i) above

$$0\ 05\ (49-40)\cdot 1{,}30 = 0{,}59 \text{ US\$ per } m^3$$

and in case (ii) above with

$$0\ 05\ (55-40)\cdot 1{,}30 = 0{,}98 \text{ US\$ per } m^3$$

This cost estimate can then be used for assessment of the profitability of increasing the mill capacity, where the economy of scale most likely will supercede the increased transport costs. The result can also be used for developing a price policy for purchasing sawtimber from private forest owners in the various belts of the supply area.

3.5 LINES SERVING AREAS

3.5.1. Road Spacing. In Chapter 2 a model for optimizing road densities was presented. It was further suggested that the spatial modifications of the model to real life road nets best is made with two empirically derived coefficients, T and V (page 66). Some very important considerations must then be made in deriving the other numeraires that enter the formulas. These are:

road cost - for construction and maintenance

variable transport cost - for all transports to and from the roads

quantity of transport

These will be discussed at some depth as an incorrect computation of these numeraires will give misleading results.

A distinction should now be made between the crosswise and lengthwise functions of the road.

These two functions are independent. A high standard of a road does not improve or influence the crosswise function, as long as a simple road is trafficable during the required periods of the year. We call such a road a minimum standard road. As the formula is based on the marginal substitution between the cost of roads and of crosswise transport, the parameter values should be:

for the road factor - the cost per unit length of a minimum standdard road, in principle for the last meter of a dead end road

for the transport cost - the cost per unit length for all crosswise transports of timber, labour, material etc. The computation of the cost of walking for the workers from the road to the work place in the paragraph 3.2.3, p. 98-100 is made in order to aggregate into one numeraire the cost of timber transport and of labour movements.

The next step is to consider the time horizon - the planning period for the model. All three parameters - road, transport and quantity - are periodic. The principle shall be to find the planning period for which the formula takes a minimum value, provided this period will not be so long that technical obsolescence (e.g. for the roads) appears.

During the planning period the cost of roads and transport may change due to technical developments. However, in most cases it is reasonable to assume that these cost changes will be of the same magnitude - be parallel - for roads and transport. In such a case the ratio of road cost to transport cost will remain the same and problems related to technical developments can be discarded. If there is valid evidence that these two cost items are likely to develope differently, then estimates of this difference should be made and the numeraires accordingly corrected. However, there are no valid reason for assuming such changes in the long term. The road cost consists of two components - construction and maintenance. Maintenance costs are recurrent at intervals during the planning period. Usually they are computed as an annual cost per unit length of the road and they can be capitalized to a present worth for the planning period and aggregated with the construction cost, using the formula

$$RC_n = CC + \frac{mc}{(1+r)^n} \qquad (3.13)$$

RC_n = road cost (present worth) for a planning period of n years

CC = road construction cost

mc = annual maintenance cost

r = rate of interest

n = planning period, years

In general, the maintenance cost for roads are correlated to its construction cost. "mc" often is in the range of 3-7% of "CC". The lower figures applies to situations with moderate rainfall, easy or rolling terrain and suitable road building material whereas the higher figures may apply to more severe conditions with heavy rainfalls, steep terrain or poor road bed material.

Finally, the transport quantity for the planning period shall be estimated. As a starting point it is assumed that the road is built just prior to an harvesting operation, preferably over the whole drainage area. Based on the inventory of the forests and the management plan, the quantities to be harvested at different points in time are estimated and capitalized to a present value of the quantity, using the formula

$$Q_i = r \frac{q_i}{(1+r)^i} \tag{3.14}$$

Q_i = present value of quantities per unit area harvested at all i years during the period

q_i = quantity harvested per unit area at the year i

r = rate of interest

The estimate should preferably be made for some alternative lengths of the planning period. The one to be chosen shall have the lowest value of the quotient $RC_n : Q_i$ as this leads to the minimum aggregate cost of roads and transports at the required rate of interest. The outcome will depend on the age distribution of the forests. In the Appendix to Chapter 3 an example on the assessment of the quantity (Q_i) is demonstrated. The conclusion of that estimate is the following:

If more than half of the drainage area consists of mature stands to be clear cut, the optimal period is only one or a few years and we need not concern ourselves about costs beyond the immediate logging operation. If the quantities fall out more evenly distributed in time, then it may be necessary to extend the planning period a few decades, but generally not longer than up to 40 years.

The final result of a road spacing exercise according to this procedure is optimal spacing over the whole area of roads built to minimum standard. It shall be followed by an estimate of which sections of that net that shall be upgraded to higher standards, using the formula derived in the Chapter 2.

3.5.2 Road Standard. The same considerations with regard to the impact of the length of the planning period as for the spacing estimate need to be made for the assessment of optimal road standards.

The key problem for the determination of road standard is the correlation between road cost and the variable transport cost on the road, which can be assumed to be: high standard - high road cost - high velocity of the vehicles - low variable transport cost.

Many situation factors have influence on the construction costs of forest roads of different standards. Standards for forest roads are often specified in terms of the average velocity of timber trucks travelling on the road. Technical specifications for the road needed to allow these speeds of travel are then established. Such road specifications include such factors as width of road, minimum radius of horizontal and vertical curves, maximum permitted gradients etc. Such specifications, often locally adapted, can be found in road construction handbooks, in instructions from authorities and enterprises and shall not be reviewed here. As a rule, road standards are specified in four or more classes and we are now interested in the corresponding costs. A generalized cost model for roads was published by Bendz (1974) with the formula

$$RC_i = 230 + 17 \cdot SL + 660 \cdot ST_i + 30 \cdot SL \cdot ST_i \qquad (3.15)$$

RC_i = construction cost in US$ per km for road standard i

SL = the inclination in percent of the major slopes (> 50 m) of the hillsides

ST_i = the road standard, which takes the following values

0 for trails for wheeled skidders and jeeps
1 for secondary feeder roads
2 for primary feeder roads
3 for main and access roads

The formula is shown in the figure 3.3. (This formula is an example of cost interdependency between two variables, SL and ST_i, reflected in this case by introducing the term $SL \cdot ST_i$.)

The authorities of the public roads maintain technical specifications on the design of roads of different standards. These specifications include figures on the maximal dimensions of vehicles permitted to travel on the public roads. For occasional single transports of bulky or heavy goods special permits may be given exceeding these figures.

In most cases considerable milage of a truck haul takes place on a public road. In such situations it is of little use to build private forest roads to higher carrying capacity than what is allowed on the public roads. However, some outer sections of a forest road net may be built to a standard allowing full loads only under the favourable seasons of the year but generally speaking the road net should be built to the same standards as the public highways with regard to weight, width and length of the vehicle but not necessarily to speed and safety.

As exemplified in table 3.3, p. 113 it may sometimes be feasible in afforestation schemes to build roads of lower standards during the establishment phase (Est. in table 3.3) to be upgraded when the logging commences (Log in table 3.3).

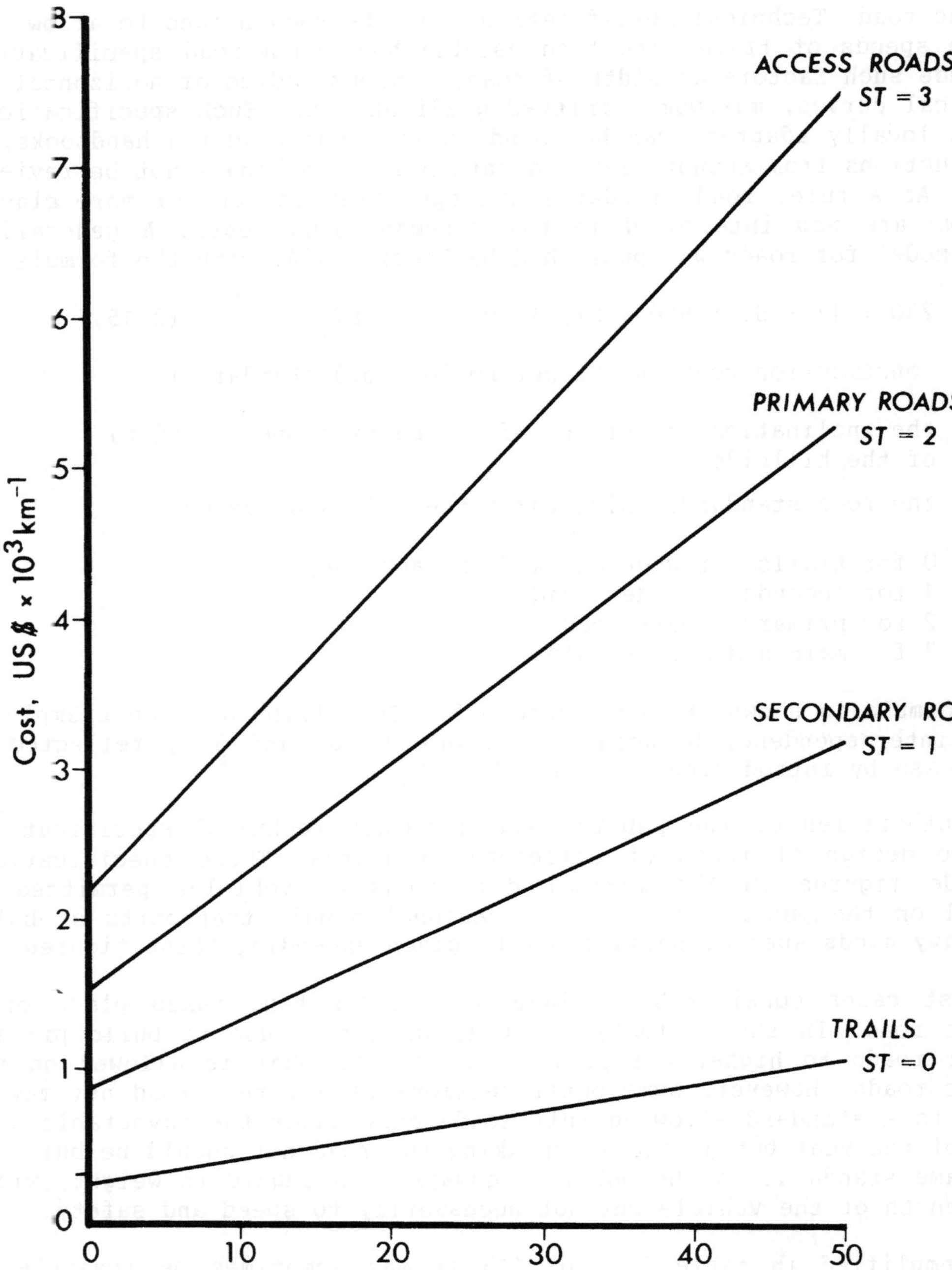

Figure 3.3. Direct construction cost of forest roads (Bendz, 1974)
Note: Cost level in the early 70's.

In the Chapter 2 we developed a formula for optimum road standard based on the following correlation

$$RC_i = a + b \cdot v_i \tag{3.16}$$

RC = road cost for road standard i

v_i = average speed of vehicle on road standard i, for roundtrip

a, b = technical constants

The fact that forest roads in addition to timber transport also carry workers and goods of various kind and also may provide other benefits for which road standard is important raises the problem if and how these other functions shall be included in an optimization estimate. Generally applicable guidelines can with our present knowledge not be given, mainly for two reasons. The specific lengthwise function of a forest road net varies widely from case to case and secondly, the empirical data base for deriving valid and generally applicable coefficients for the mathematical model for other functions than for timber transport is very weak. It is therefore suggested that this model be used to optimize road standard for timber transport only. This will result in an upgrading program of limited scale for which the cost of upgrading with certainty is recovered by cheaper timber transports. If a further upgrading is justified for other additional functions, this should be clarified in separate assessments or simply by logical reasoning.

Another approach for establishing an upgrading plan is a purely empirical one: to record the break down on road standards of already existing road nets in similar situations. It is reasonable to assume that such a road net during years of use was adjusted to the specific demands and requirements of the users. It may consequently be close to an optimum design. An example is given in the table 3.3 for a break down of a road net for trucks on three road standards in three terrain types.

Table 3.3 Example of breakdown on road standards for forest road nets in three terrain types, in percent

	Terrain type					
	Undulated		Steep		Very steep	
Road standard	Est.	Log	Est.	Log	Est.	Log
Suggested road density, m/hectare	17,5	36	18	26	18	32
"- miles per sq. miles	2,8	5,6	2,9	4,2	2,9	5,02
Artery, main roads, percent	10	20	15	20	20	20
Primary roads, percent	30	30	35	35	40	40
Secondary roads, percent	60	50	50	45	40	40
	100%	100%	100%	100%	100%	100%

(From Sundberg, 1973. Note: The example refers to a pine plantation project in a tropical/subtropical area. Est. = Period of establishment and tending up to the time when logging commences = Log.)

As suggested in the table 3.3 the lowest road standard often applies to 40-60% of the whole road net, sometimes even more whereas the best standard is limited to around 10-20% or less. The pattern of the road net also bears on the break down. If through roads are common and dead end roads less frequent, the ratio of the lowest and the highest standard are likely to shrink.

3.5.3 Forest Management and Road Planning. The previous review of methodology and applications to real life situations of forest road planning suggests the following procedure.

Establish first the optimum spacing (or road density) for the whole area of roads built to minimum road standard. Then assess what fractions of that road net that shall be upgraded to higher road standards. In these estimates the age distribution of the forest stands, the volumes to be harvested according to management plans must be considered and an appropriate planning period chosen. The dynamic nature of a road plan must be observed, a consequence of periodic variations in time of harvested volumes, as a rule caused by an uneven age distribution of the forest stands. The road plan may therefore contain a successive extension of the road net, as logging operations extend to new areas.

It may even include the closing down of parts of an already built road net in cases when - after massive clearcutting - only minor logging is foreseen next few decades.

The problem of establishing correctly the planning period is often influenced by the ownership structure. If all forests within the drainage area are owned by one subject, he should log the whole area immediately after the construction of the roads and not return until after a treatment interval as prescribed in the management plan. In such cases, it is as a rule best to choose the shortest possible planning period. If on the other hand the forests are owned by a large number of forest farmers, logging operations are likely to be more evenly spread out in time and the roads are required to be in good shape every year. Then, most likely, a planning period of 20-40 years should be used. These arguments clearly demonstrate that roads are an integer part of the whole forest management system. The mutual interdependency between forest management and roading shall be observed by all staff members.

It can be mentioned that mathematical models have been worked out which simultaneously optimize spacing and standard (Larsson, 1959, Larsson and Rydstern, 1968). These models contradict the conclusions advocated in this text in one aspect: road spacing can be substituted by road standard. However, estimates according to both methods give results which do not differ much. In any case, the mathematics of such simultaneous optimization is much more complex and tend to obscure the main purpose of this text - to demonstrate the structure of logging and roading problems and how they can be addressed with analytic tools.

3.6 COSTING OF MACHINES

Methods for calculating machine costs are described in most engineering handbooks. Therefore, only some specific applications for forest machines shall be presented.

3.6.1 Thumb-rule Proceedures. Samset (1985) suggests that the hourly cost of a machine can be determined with the simple formula

$$C_d = 2 \cdot A \cdot 10^{-3} + 5 \qquad (3.17)$$

where

C_d = Machine cost US$ per day, exclusive operator

A = Purchase price of the machine, US$

Another simple method (Sundberg, 1982) suggests that the machine cost can be determined on the basis of its fuel consumption. (See further Chapter 4.) The supporting theory is that fuel consumption as well as the need for repair, maintenance and depreciation are functions of the actual volume of work performed and consequently all costs of a machine are directly related to its fuel consumption. In the Appendix C to Chapter 4 "Work Load Factor - Power Rating" some additional views are provided on the assessment of the fuel consumption. See also Appendix B to Chapter 1.

3.6.2 Detailed Costing. For very costly and complex machines, such as helicopters, separate estimates are often made for different mechanical components, especially if replacements, servicing and repair requirements are very different for various machine components and components have different life expectancies. This is seldom needed for forest machines, except that items like saw chains of a power saw or tires of a truck are subject to separate cost estimates and not included in the aggregate unspecified repair and maintenance cost.

3.6.3 Standard Costing Methods. Some forest jobs put widely varying demands on a machine during different phases of the work and consequently generate a different cost per unit time. Typical are trucks (travelling and idling while being loaded and unloaded) and sometimes loaders. In such cases different costs per unit time may be derived to represent the true cost for each time sequence. An application to such cases will be outlined below, using a standard costing method.

The fixed costs of the machine are first aggregated per shift hours and then the additional, variable costs imposed by the actual work are allocated on those hours of the shift when the machine actually performs work, e.g. the travelling time of a truck, (see the form on next page). The cost per work hour will then be the sum of fixed and variable costs and for idle hours of the shift the fixed hourly cost only.

MACHINE OPERATING COST ESTIMATE

Machine: Description ____

Gross kW______________ Delivered cost______________

Life in years______ Hours (days): per year____ life________

Fuel: Type____________________________ Price per litre___________

Tires: Size____________ Type_____________ Number_________________

Cost of replacement set___________________________________

Operator: Rate per hour(day)__________ Fringe benefits____________%

Cost component (fixed) Cost per hour (day)

(a) Depreciation = $\frac{\text{delivered cost} \cdot 0,9}{\text{life in hours}}$

(b) Interest = $\frac{\text{delivered cost} \cdot 0,60 \cdot \text{interest rate}}{\text{average hours per year}}$ ____________

(c) Insurance = $\frac{\text{delivered cost} \cdot 0,60 \cdot 0,3}{\text{average hours per year}}$ ____________

(d) Taxes = $\frac{\text{annual tax amount}}{\text{average hours per year}}$ ____________

(e) Op. labour = $\frac{\text{Labour cost for period} \cdot (1+f)}{\text{machine hours in period}}$ ____________

where f = cost of labour fringe benefits

in % of direct labour cost

Cost component (variable)	Fixed cost per work hour
(f) Fuel = gkW · X · CL where gkW = gross engine power CL = fuel cost per litre in dollars X = 0,18 for diesel fuel, 0,25 for gasoline	______
(g) Oil and grease = $\frac{\text{gkW} \cdot \text{X}}{100}$ where X = 0,20 for tractors, skidders, front-end loaders and trucks X = 0,30 for feller-bunchers and knuckle-boom loaders X = 0,50 for processors, harvesters and forwarders	______
(h) Servicing and repairs (*) = $\frac{\text{Delivered cost}}{\text{life in hours**}}$ * include tires except for hauling rigs ** lifetime travelling hours in case of hauling rigs (around 2000 h for tires)	______
(i) Tires for hauling rigs = 0,0006 · CST where CST = cost of set of replacement tires	______
Variable cost per work hour	______
Total cost per work hour (fixed + variable)	______

4. ENERGY ANALYSIS

4.1 ENERGY IN FORESTRY - AN INTRODUCTION

Energy is significant for forestry from two different points of view. A tree crop captures solar energy, usually regarded as free energy and stores the energy in the form of biomass. A considerable part of the biomass of the tree crop is being used to satisfy human needs. The second aspect on energy in forestry is that for the management, harvesting and conversion of the biomass to marketable or useful products energy is used. This input is not in the form of free energy but as energy bound or embodied in petrol, machines, fertilizer etc. Forest labour requires energy in many forms to be able to perform work as do the animals used in forest work (horses, oxens, mules etc.).

Solar energy is regarded as free energy for which we do not have to pay a price. However, in order to capture it and make use of it we need an area that intercepts that energy and provides the base for tree growth and this is exactly what forestry is about, to exploit the land for energy conservation. The forest owner or whoever has tenure of this land is the only person who can make use of the solar energy falling on his property so in a way part of the soil rent for the land can be looked upon as a price for solar energy. That is why land was basis for wealth before industrialization. Forest land without solar energy is useless for the sustained production for which the soil rent is relevant.

Energy bound in fuels and capital equipment always carries a dollar cost. Such energy in production will eventually be used up and dissipated as waste heat of such low temperature that people no longer can use it. Forestry then can be regarded as production processes in which energy is transformed. One process is the growing of forest biomass on land in a ecosystem in which solar energy is transformed and concentrated into usable material. The second process is the use of stored energy for manipulating this ecosystem, to harvest and convert the material to products demanded by man. The energy is transformed into work and is eventually lost as waste heat.

The purpose of this chapter will be to study the production processes with regard to their energy efficiency because high efficiency must always be desirable. Efficiency in production traditionally is analysed from the point of view of time or money. During the last two decades a new science has emerged using energy as the measuring stick. These analyses, no doubt, give new insights into production, into the mechanisms and factor relations in production, which promise to give researchers new insights and production manager a new tool of achieving the objectives of the production process and also guidelines to improve it. We shall mainly deal with the second energy transformation in forestry in which bound energy is put into the production process and used up. The first phase, the capturing of solar energy in the photosyntheses in the ecosystem will be discussed only marginally.

The dramatic increases in the cost of energy in the 1970's very clearly disclosed the profound role of energy in our economies, for example in the effects on national wealth, inflation, labour productivity and so on. We now can better understand the underlying reasons for and the economic effects of the rapid mechanization of forest work in the 1950's and 1960's - increasingly cheaper energy in relation to other inputs, e.g. labour. The disruption of the national economics in almost all countries in the 1970's and early 1980's no doubt was caused from the sudden and repeated price shocks on fuels to which it was impossible for national economies to adjust quickly. The conclusion of these events are that as energy plays such distinct role in all economic activity it must be meaningful to put forestry in an energy perspective. "Surely the time has come to understand exactly how energy enters the economic process" (Slesser 1978, p. 3).

4.2 THERMODYNAMICS

Energy is an abstraction, like money, but its effects are very clear. We know that when heat or work are put into or taken out of a system, that system ends up in a different state. Some property of the system has to account for the difference. In thermodynamics it can be expressed as follows (Slesser 1978, p. 11).

$$E_f - E_i = \Delta E = J\Sigma Q - \Sigma W \qquad (4.1)$$

where

ΔE = change in energy

E_i = initial content of energy

E_f = final content of energy

Q = amount of heat taken in by the system

W = work done by the system

J = a constant, the heat equivalent of work

The first law of thermodynamics states that the energy can be neither created nor destroyed but is always conserved in the sense that it cannot be thrown away. A given amount of work is equal to a precise amount of heat. The measure of work is kilowatt-hour and the measure of heat is joule or calory.

Table 4.1 Heat and work conversion factors

Heat		Work
megajoules (10^6 joules)	kilocalories (10^3 calories)	kilowatt-hours
1	239	0,2778
$4,184 \cdot 10^{-3}$	1	$1,115 \cdot 10^{-3}$
3,6	860,45	1

Work can be transformed into heat. If we rub our hands the skin is warmed up. The reverse process, to change heat into work is crucial to the understanding of the role of energy in production. Heat can spontaneously flow from a hot place to a cold one, never in the reverse direction. But we can make heat flow from a cold place to a hot one - that is what a refrigirator does but it does not spontaneously, for it requires energy to make the heat flow in the "wrong" direction. The second law of thermodynamics states that the energy required to make heat flow from a hot place to a cold one always is more than what is yielded in the spontaneous reverse process. This may seem to be a simple theorem but the implications are far-reaching. The fraction of a quantity of heat that can be usefully turned into work is given by

$$(T - T_0) : T \tag{4.2}$$

T = temperature of heat supply

T_0 = temperature of the "sink" into which the used-up heat goes

The temperature has to be turned into absolute temperature units by adding 273 to the centigrade scale (Celsius). The ratio (4.2) above is called the Carnot efficiency after a French engineer who first formulated the theorem. Using -273°C as sink explains the strange notion that there is actually more energy in the water of the North Sea than in the oil known to be under it, no less than 500 times (Slesser, 1978). But the important, practical conclusion of Carnot's theorem is that the efficiency in the use of that energy increases the more the upper temperature, T, is raised. Fuels which can be burned at high temperatures (e.g. oil) can produce more work per unit thermal heat than fuels burned at lower temperature (e.g. wood, peat). Entropy is a measure of order and can be defined as a measure of the unavailability of a system's thermal energy for conversion into mechanical work. The water in the North Sea has high entropy, the oil under it low. The term "entropy" was coined by the German scientist Clasius who draw the conclusion:

Die Entropie der Welt strebt einem Maximum zu.
The Entropy of the world grows to a maximum.

This doomsday expression is actually telling us that all energy in the final end will be uniformly dissipated as waste heat in the earth, unavailable for any production purposes for mankind. The Limits of Growth by the Club of Rome can be seen as act I in this scenario. Solar energy is the only counteracting source and forestry can contribute by improving the production of low entropy materia - biomass - and by making this materia available to meet human needs with highest possible efficiency. The past developments of forestry and forestry sciences can be understood in terms of thermodynamics. The thermodynamic relation between heat content H and energy content E is

$$\Delta H = \Delta E + P\Delta v \quad (4.3)$$

P = pressure
Δv = change in volume of the system

Δv is normally small and is often neglected in which case heat content and energy content become equal concepts. Thus the concept energy is often used as heat content relative to a standard state of pressure one bar and zero degrees centigrades.

4.3 PHOTOSYNTHESIS

Almost all improvements in forest management conform with the objectives improving yields from photosynthesis and of harvesting and converting the biomass with a high level of efficiency. Energy plantations are one such effort, which is directed towards energy production. Traditional silviculture for the production of sawlogs and pulpwood also aims towards the same goals.

The flow of energy from the sun to the planet earth is immense. It is estimated (King Hubert, 1971) that it is 21 000 times more than the total commercial energy consumption. But only some 0,3 - 0,4% is captured by biosystems on land and in water (Slesser 1978, p. 74). Photosynthetic efficiency is the ratio of the solar energy captured by plants to the incoming solar energy.

The photosynthesic equation can be written as follows

$$6CO_2 + 12H_2O \xrightarrow{\text{sunlight}} 6_6H_{12}O_6 + 6O_2 + 6H_2O \quad (4.4)$$

Carbon-dioxid — Water — Glucose — Oxygen — Water

Some tropical plants - maize, sorghum, sugar cane and some grasses, the C_4 plants, have a somewhat higher photosynthetic efficiency than the C_3 plants to which trees belong.

The photosynthetic efficiency in traditionally managed forests in sustained yield forestry is about 0,2% in the form of presently used products, while the energy conversion in intensive cultures under optimal field conditions can reach values ranging from 3 to 5% during the growing season. Theoretically it should be possible to increase the biomass production from forest land some 10 to 20 times and that is exactly what energy plantations are about. In, figure 4.1, some expected plant yields are given. In practice we have thus far being able to increase efficiency only to less than half of this potential.

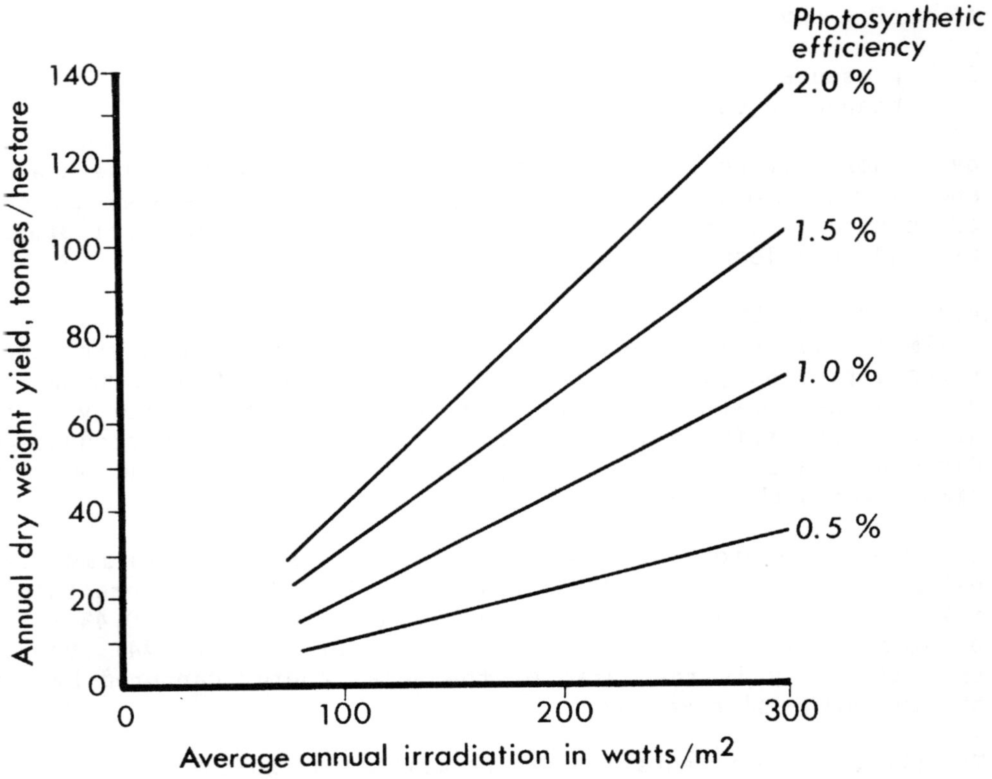

Figure 4.1. Expected annual plant yields as a function of annual solar irradiation at various photosynthetic efficiencies. (From Hall, 1977)

The table 4.2 (from Hall, 1977) shows the yields which can be expected under various sunlight intensities for some crops.

The efficiency of the photosynthesis is now being studied intensively by many research institutions throughout the world in attempts to achieve increasing yields of biomass for energy supply from a sustained and renewable system. In practice, the plants require a steady supply of water to grow well and this is often a critical and costly item, especially in dry zones. Even short periods of insufficiency of water may reduce growth for much longer time.

This research on biomass for energy also involves genetic manipulation of the plants and other schemes for improvements of the photosynthesis. There are many who believe that there are great potentials for success.

Table 4.2 Average-to-good annual yields of dry matter production

	Tonne hectare^{-1} year^{-1}	g m^{-2}day^{-1}	Photosynthetic efficiency (% of total radiation)
Tropical			
Napier grass	88	24	1,6
Sugar cane	66	18	1,2
Reed swamp	59	16	1,1
Temperate (Europe)			
Perennial crops	29	8	1,0
Annual crops	22	6	0,8
Grasslands	22	6	0,8
Evergreen forest	22	6	0,8
Decidous forest	15	4	0,6
Savanne	11	3	
Desert	1	0,3	0,02

4.4 PROCESS ANALYSIS

The role of energy in the economics of nations, industries and humans has been subject to a large number of studies in the past few decades. In 1974 the IFIAS workshop laid down a precedural methodology for energy analysis*). Such analysis was defined as: Determination of the energy sequestered in the process of making a good or service within the framework of an agreed set of conditions. We use the term industrial energy to identify fossil.

Numerous studies are made on the interrelation between energy use and level of economic output. The strong correlation between GNP and energy use is established in many nations and often used for the forecasting of future needs of energy. We shall not attempt to review this type of studies on the macro level but refer to Slesser (1978) and Smil and Kuz (1976) for European countries and Schurr and Netschert (1980) and Cleveland et al (1984) for the USA. It is important to note that although these has been a strong correlation between energy use and economic activity this does not imply an "iron law", and it is an important goal for all nations to try to increase their efficiency of energy use. We shall concentrate on what IFIAS termed Process Energy Requirement, PER, which mainly deals with one sector of a national economy or specific work operations for the purpose of comparing one technology of production with another.

---

*)IFIAS = International Federation of Institutes for Advanced Study, Stockholm

Two cases will be shown as examples. First, figure 4.2 shows the flow of energy and material in the forest sector in the ECE countries.

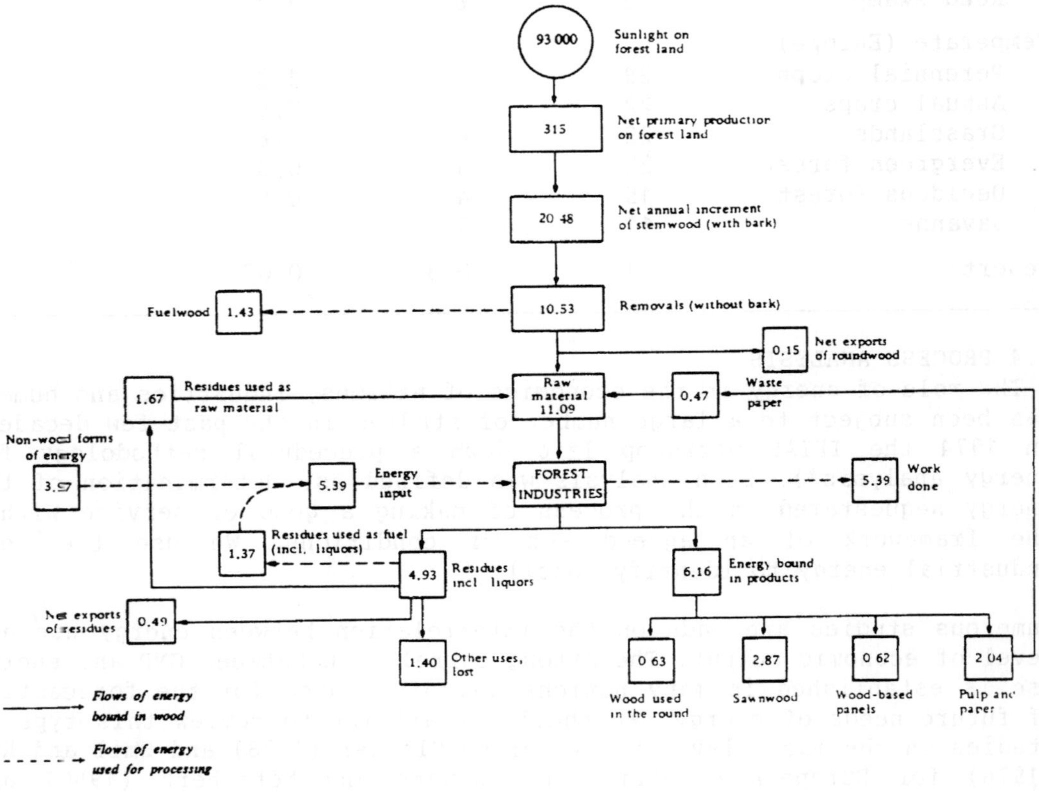

Figure 4.2. ECE region: energy flows in the forest industries in the mid 1970's (10^6TJ).

The removals of wood from the forests in the ECE Region had an energy value of 10,53 $\cdot 10^6$TJ and the outputs from the industries (after deducting 1,43 for fuelwood and 0,15 for exports of roundwood) had an energy value of 8,08 $\cdot$ 10^6TJ.

The flow of matter and energy in the work process is shown in simple form in the figure 4.3.

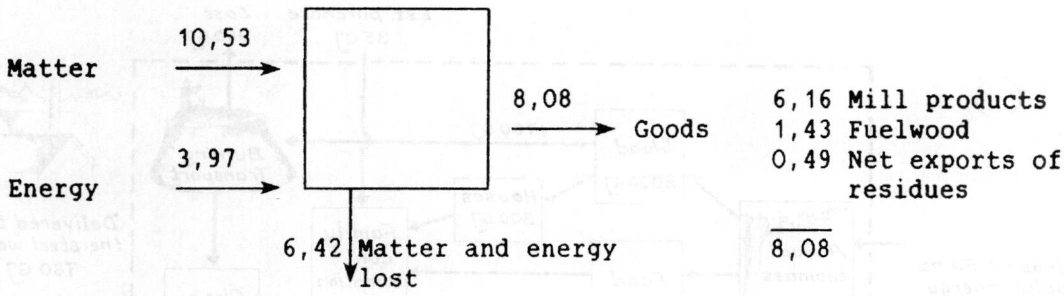

Figure 4.3. The flow of energy and matter in the forest sector in the ECE region. (10^6 TJ). Simplified version of figure 4.2.

In chapter 1.1 p. 1-9 the operations of a forest enterprise in Sweden was reviewed, including a graph, figure 1.4, p. 9 an input/output analysis of energy. This enterprise can be looked upon as a subsystem of the ECE total region. For the enterprise the output/input ratio was 20 TJ out per TJ in while for the ECE system, which includes also the manufacture of mill products, the ratio is around 2 TJ out per TJ in. It can be observed in figure 4.2 that the industries use to a considerable extent residues for energy (1,37 out of 5,39 · 10^6TJ), mainly liquors in pulp mills.

A second example on forest energy is an attempt to describe a charcoal burner in Sweden some 150 years ago, figure 4.4. It is estimated (Sundberg, 1985) that this man and his family used only 35 GJ of external energy but delivered to the market around 760 GJ in the form of charcoal per annum, an output/input ratio of 22.

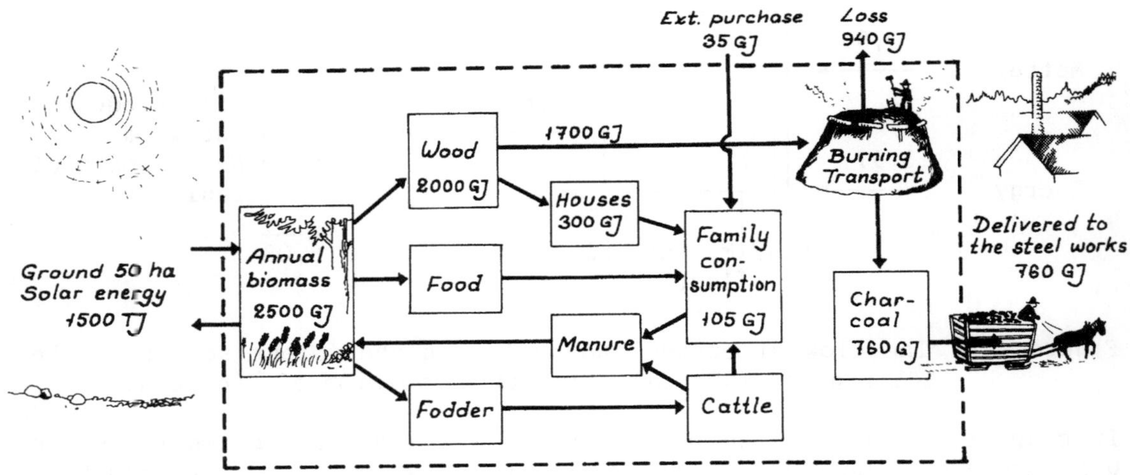

Figure 4.4. Energy system for a charcoal burner at Basttjärn and his family, who used 2 hectares of tilled ground, 8 hectares of pasture and 40 hectares of forest. I terrajoule (TJ) = 1 000 gigajoule (GJ) = 10^{12} joule. 1 GJ = 278 kWh. 1 tonne of oil is the equivalent of about 42 GJ.

Forestry and agriculture deliver energy in the form of wood and food. The more intense the cultivation the more input of energy is generally required as shown in table 4.3.

Table 4.3 Inputs of energy and outputs

	INPUT GJ per hectare	OUTPUT Harvest, kg grain per hectare
Agriculture: world1)	7,9	1 821
Developing countries1)	2,2	1 255
Developed countries1)	24,8	3 100
Forestry: Sweden2)	1,0	around 4 cubic meter of wood
Energy plantations (intense)3)	57,1	78,2 dry tonne biomass

Source: 1) FAO; 2) Sundberg, 1982; 3) Inman

The energy factor in agricultural production has been studied intensively. For a forester, who in a sense competes with agriculturists for land, some indicative figures are of interest.

Leach has compiled energy ratios (in this case food energy out/ industrial energy in) for food production. 'Energy out' in relation to 'Energy in' is around 0,2-0,5 for agriculture as a whole where as for cereals the ratio is around 2-4. Highest ratios, 15-40 is obtained in subsistence agriculture in the tropics, in which no machines or fertilizers are used.

An interesting result in Leach's study is that the energy input per £ value of the output was almost the same (around 200 MJ/£) for such different production lines as dairy, cattle, sheep, pig, poultry and cereals. This supports Costanza's idea, referred to later in this chapter, that markets operate to use energy most efficiently.

Energy content fuels have somewhat varying proportions of carbon, hydrogen, sulphur etc and therefore their heat content may vary accordingly. In the Appendix A to Chapter 4 the heat content of some energy fuels are given.

We are mainly interested here in the content of work contained in these fuels, mainly for fuels used in combustion engines.

Table 4.4 Energy content in some fuels (l = litre)

	kWh		MJ		Specific
	l	kg	l	kg	weight
Diesel oil	9,88	11,86	35,59	42,71	0,83
Benzin	9,72	11,98	31,40	43,13	0,73
Kerosene	9,71	11,98	34,96	43,13	0,81
Wood, dry	-	5,33	-	19,2	-

Only a fraction of this work content is transformed to useful work in a machine.

In studies relating GNP to energy use researchers have established quality values or economic energy values to reflect the fact that different fuels have different economic effectiveness or utility. For example, an electric motor can be used more precisely and efficiently than a steam engine. These show a considerable variation but some approximate averages are given below.

Table 4.5 Approx. economic energy values for fuels

Coal	1
Oil and Gas	around 2
Electricity	2,5-3

As these values have their origin in economics, they shall not be mixed up with the efficiency of a machine transforming energy to work.

4.5 SYSTEM BOUNDARIES

Energy analyses are concerned primarily with the accounting of energy in economic production. A production process or system is selected and the application of energy to matter and its flow through the system is mapped and put in relation to the output of goods and services. Both methodologies and aims are analogous to economics but the numeraire in energy studies is heat or work and in economics money. Several attempts are being made to bridge the gap between energy and economics which later will be exemplified and discussed.

A fundamental problem in energy analyses is to decide upon a logical and consistent system boundary. In particular the upstream flow of energy in the system can be delimited in various way. Depending on how much of the upstream flows are included, the figures on energy input can show variations. Thus, in comparisons of results from different studies the delimitation of the system must always be kept in mind. The purpose of the IFIAS workshop in 1974 was to establish some standard procedures to make results comparable and consistent. Two extreme examples can illustrate the problem. IFIAS coined GER, short for Gross Energy Requirements which is the sum of all the energy resources which has to be sequestered from the earth's non-renewable resources of energy in order to deliver a good or service to the point of interest (Slesser, 1978).

The objective of a good energy analysis is to determine how much energy is really needed to produce the object in question. In practice decisions as to where to trancate the analysis usually can be made without much trouble. For example, the energy required to make a saw certainly includes the energy used in the saw factory and to mine the iron, and even in the shovel used to mine the iron. It can include the energy used to grow the food of the shoveler. But there is no reason to include the food need to feed the fist caveman who figured out how to smelt iron.

Oil has a calorific heat (energy) content, but energy also is used up to explore an oilfield, to drill, pump, transport, refine etc the crude oil until it is used in production. These energies are called the embodied energy of oil. The GER includes all these energies. Choosing a "small" production process, like logging and transport, we may use only the calorific heat value of the fuels, generally termed direct energy as energy input. But this approach neglects large quantities of energy which enter the system in other ways and which are used up in the production process and largely dissipated, as high entropy, heat. Examples include energy used to manufacture the forestry machines, energy used to build and maintain the roads and other installations, energy for transport of labour and so on. This can be illustrated in a graph on an energy analysis of a transport system.

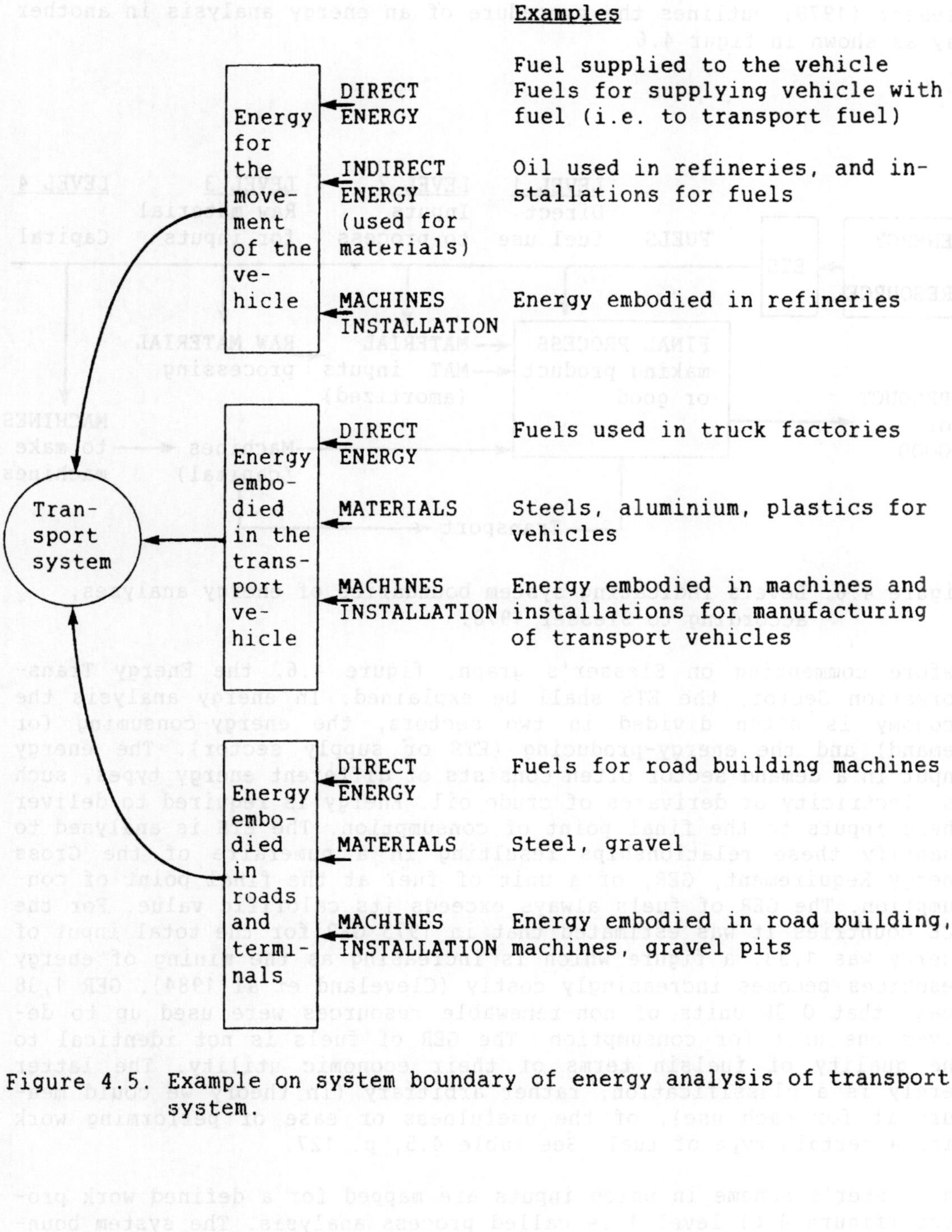

Figure 4.5. Example on system boundary of energy analysis of transport system.

Slesser (1978) outlines the procedure of an energy analysis in another way as shown in figur 4.6.

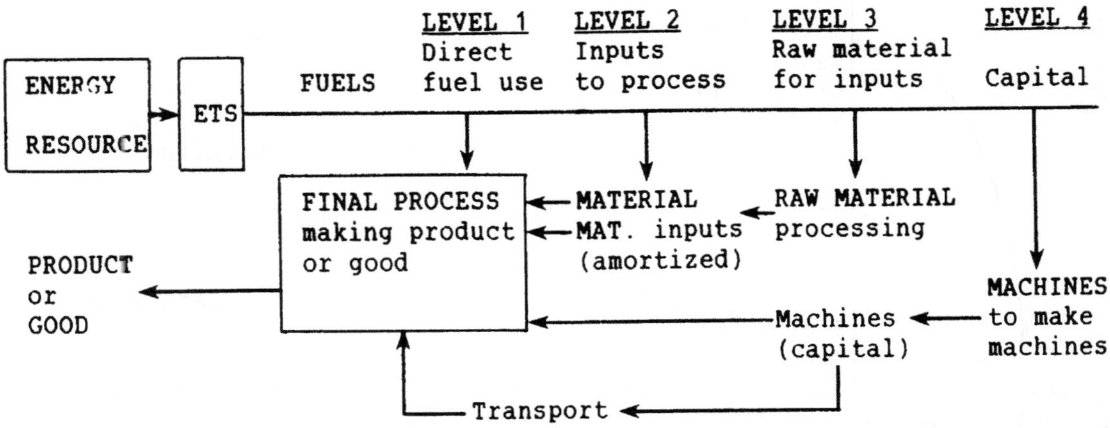

Figure 4.6. Levels indicating system boundaries of energy analyses, according to Slesser 1978.

Before commenting on Slesser's graph, figure 4.6, the Energy Transformation Sector, the ETS shall be explained. In energy analysis the economy is often divided in two sectors, the energy-consuming (or demand) and the energy-producing (ETS or supply sector). The energy input in a demand sector often consists of different energy types, such as electricity or derivates of crude oil. Energy is required to deliver these inputs to the final point of consumption. The ETS is analysed to quantify these relationships resulting in a numeraire of the Gross Energy Requirement, GER, of a unit of fuel at the final point of consumption. The GER of fuels always exceeds its calorific value. For the ECE countries it was estimated that in 1973 GER for the total input of energy was 1,38, a figure which is increasing as the mining of energy resources becomes increasingly costly (Cleveland et al 1984). GER 1,38 means that 0,38 units of non-renewable resources were used up to deliver one unit for consumption. The GER of fuels is not identical to the quality of fuelsin terms of their economic utility. The latter merely is a classification, rather arbitrary (in theory we could measure it for each use), of the usefulness or ease of performing work with a certain type of fuel. See table 4.5, p. 127.

In Slesser's scheme in which inputs are mapped for a defined work process (figure 4.6) level 1 is called process analysis. The system boundaries for a process analysis can be drawn up at will, depending largely on the purpose of the study. In the most simple analysis of forest operations, where practically fuel inputs consist only of petroleum products, levels 2-4 may often be ignored and kept outside the system boundaries. Perhaps this can be justified in cases when the same type of fuels are used in equal proportions for the work alternatives to be compared. But as soon as the proportions of fuel types are altered or the impact on depletion of energy resources is relevant for the study the GER of the fuels should be used. Especially electricity has a large impact in this respect.

Forestry is a particular sector in economics in the context of energy as it is both consumies and produces energy. Before the industrial revolution some centuries ago energy production was often dominant and that is still the case in many developing countries. Energy plantations belongs clearly to the energy producing sector. In developed countries traditional industrial forestry for sawlogs and pulpwood dominates and the energy producing component often is small although it is being revived slowly in some countries. Surprisingly, Koning and Skog (Biomass, Vol. 12, No. 1, 1987) state, that the end use of most wood in the U.S. is not lumber or pulp and paper but feed for energy. In 1983 no less than 128,4 MM cords of wood was used as feed for energy, which is considerably more than the wood content of the industrial products.

There is also another characteristic industrial forestry, the reuse of the end use products. Lumber and paper are potential energy suppliers, once their life as a product is ended. This was recognized by the IFIAS workshop which defined the Net Energy Requirement, NER, for timber. NER then is the GER for installed timber minus the energy content of it plus the energy required to make it avalable for combustion. According to Slesser the NER has not been used much, perhaps because so few complete studies on forestry have been made. The study of the forest sector in the ECE countries, shown in figure 4.2, p. 124 is incomplete in the context of NER. Calorific energy values are accounted on the end products but the energy requirements to make them available for energy production after their service life are not estimated.

When calculating GER one should include only those energies with alternate economic uses. For example the vast energy flux of the sun cannot be tapped into other useful forms of energy (except what is stored through the photosynthesis) and there is no depletion of the stock of energy as the sun is an infinite and non-depletable energy source. Thus, we do not include the flow of the sun in most of our routine GER estimates.

If solar energy becomes a source of technically useful fuels (or power) on a large scale, it will do so by taking up land. If this land could be used to produce food or wood, the trade-off is between a solar fuel or power output vs. a food or wood output. If a solar device does not take up useful land - e.g. roof top arrays, then it competes with nothing and has no meaning in terms of energy or a 'land cost'.

Another way to look at the quality of a fuel resource is to use what is called EROI = Energy Return on Investments (Cleveland et al 1984). EROI is the ratio of gross fuel extracted to economic energy required directly and indirectly to deliver the fuel to society in a useful form. It generally includes a quality correction for different fuel types, i.e. electricity is weighted by about 3 compared to coal in both numerator and denominator. The numeraires in table 4.6 are compiled (from Cleveland et al 1984).

Table 4.6 Estimates on EROI for some fuel supply technologies, USA

Process	EROI
Non renewable	
Oil and gas (domestic wellhead)	
1940'2	Discoveries >100
1970's	Production 23,0, discoveries 8,0
Coal, (mine mouth), 1970	30
Renewable	
Ethanol (sugercane)	0,8 to 1,7
" (corn)	1,3
Methanol (wood)	2,6
Firewood	30

These estimates show that, in general, it takes far less energy to process fossil fuels than similar fuels from biomass. Although biomass fuels have low ratios (except for heat, firewood) they may be improved through technology whereas fossil fuels are declining due to depletion. When they cross, perhaps sometime in the next century, biomass fuel should compete with fossile fuels.

4.6 ENERGY INPUTS IN FOREST WORK OPERATIONS

The energy balance for the forest sector in the United States in 1980 is shown in table 4.7 according to a report by Zerbe et al (1986).

Table 4.7 Energy balance in the forest sector in USA in 1980 (Zerbe et al 1986)

Growth	$6\ 859 \cdot 10^{15}$J
Annual cut	$4\ 012 \cdot 10^{15}$J
Management of the forests	$3 \cdot 10^{15}$J
Harvesting and transport	$184 \cdot 10^{15}$J
Processing to finished products	$2\ 788 \cdot 10^{15}$J

J = joule

The energy efficiency for the growing of wood (growth/management) was thus around 2 000 to 1 but varied for the US regions between 200 to 1 to 2 500 to 1. The energy efficiency for the growth and the retrieval of wood from the natural forests was 17 to 1 (or an EROI of 17).

The American Pulpwood Association in 1975 (Erickson 1983)estimated that about 9,5 litre of petroleum products are required to cut and deliver one metric tonne of wood (with a fuel equivalent of 230 litre of petroleum products).

A review of energy studies for Canadian forestry was presented by the ENFOR Project (Ash et al 1980).

The input of industrial energy in Swedish forestry in 1956 and 1972 is shown in table 4.8, expressed in litre oil equivalents (LOE) per cubic meter wood, cut and delivered to the processing mills. The corresponding input of labour is also given.

Table 4.8 Input of labour and energy in Swedish forestry in 1956 and 1972 (computed from Renborg & Uhlin 1975)

	1956	1972
Input of energy, direct and indirect, LOE/m^3	4,1	7,9
Input of labour, manyear per $m^3 \cdot 10^3$	3,35	0,64

The substitution of labour for energy during this period of rapid mechanization is clearly demonstrated. However, there were also other structural and operational changes influencing these ratios of inputs.

The distribution of the energy input on various operations are shown in figure 4.7. Logging and transport operations required about 60% of the total input, of which truck hauling was large share.

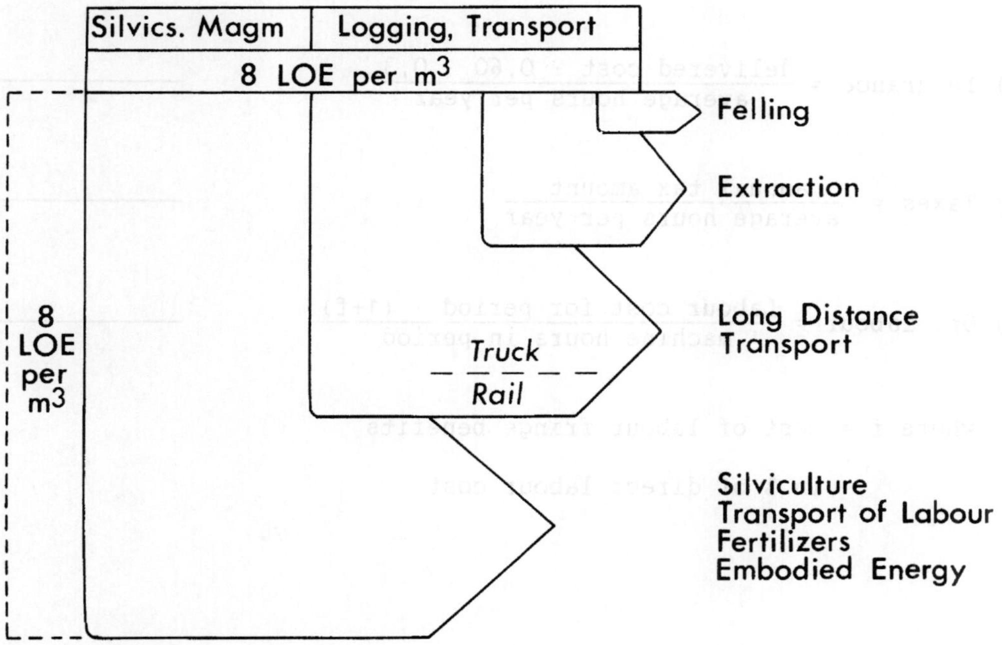

Figure 4.7. Energy consumption in Swedish forestry in 1972, expressed in litre diesel oil equivalents (LOE) per m^3 wood under bark, cut and delivered to the mills (computed from Renborg and Uhlin 1975).

MACHINE OPERATING COST ESTIMATE

Machine: Description ______________________________

Gross kW______________ Delivered cost______________

Life in years______ Hours (days): per year____ life________

Fuel: Type______________________ Price per litre__________

Tires: Size__________ Type__________ Number______________

Cost of replacement set______________________________

Operator: Rate per hour(day)________ Fringe benefits__________%

Cost component (fixed) — Cost per hour (day)

(a) Depreciation $= \dfrac{\text{delivered cost} \cdot 0{,}9}{\text{life in hours}}$ ____________

(b) Interest $= \dfrac{\text{delivered cost} \cdot 0{,}60 \cdot \text{interest rate}}{\text{average hours per year}}$ ____________

(c) Insurance $= \dfrac{\text{delivered cost} \cdot 0{,}60 \cdot 0{,}3}{\text{average hours per year}}$ ____________

(d) Taxes $= \dfrac{\text{annual tax amount}}{\text{average hours per year}}$ ____________

(e) Op. labour $= \dfrac{\text{Labour cost for period} \cdot (1+f)}{\text{machine hours in period}}$ ____________

where f = cost of labour fringe benefits

in % of direct labour cost

furnace efficiency. Note that the end product in this model is useable heat output from the furnace, not the energy content of the delivered biomass.

The operations involved as the wood passes through the system is shown in the figure 4.10. A generalized model of such an energy supply system is demonstrated in figure 4.11.

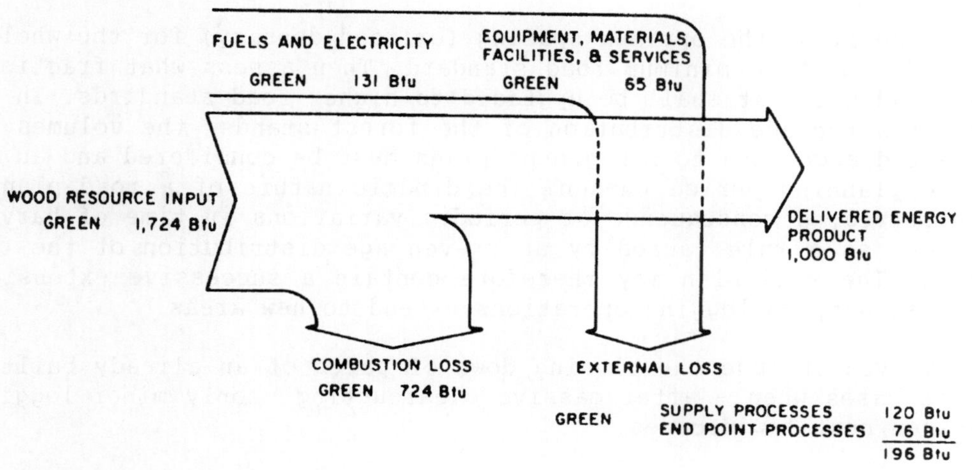

Figure 4.9. Energy balance for the entire fuelwood trajectory (fig. 4.10). Delivered energy product (1 000 BTU) - external loss (196 BTU) = net energy (804 BTU). (Baltic et al 1983)

As suggested in the table 3.3 the lowest road standard often applies to 40-60% of the whole road net, sometimes even more whereas the best standard is limited to around 10-20% or less. The pattern of the road net also bears on the break down. If through roads are common and dead end roads less frequent, the ratio of the lowest and the highest standard are likely to shrink.

3.5.3 Forest Management and Road Planning. The previous review of methodology and applications to real life situations of forest road planning suggests the following procedure.

Establish first the optimum spacing (or road density) for the whole area of roads built to minimum road standard. Then assess what fractions of that road net that shall be upgraded to higher road standards. In these estimates the age distribution of the forest stands, the volumes to be harvested according to management plans must be considered and an appropriate planning period chosen. The dynamic nature of a road plan must be observed, a consequence of periodic variations in time of harvested volumes, as a rule caused by an uneven age distribution of the forest stands. The road plan may therefore contain a successive extension of the road net, as logging operations extend to new areas.

It may even include the closing down of parts of an already built road net in cases when - after massive clearcutting - only minor logging is foreseen next few decades.

The problem of establishing correctly the planning period is often influenced by the ownership structure. If all forests within the drainage area are owned by one subject, he should log the whole area immediately after the construction of the roads and not return until after a treatment interval as prescribed in the management plan. In such cases, it is as a rule best to choose the shortest possible planning period. If on the other hand the forests are owned by a large number of forest farmers, logging operations are likely to be more evenly spread out in time and the roads are required to be in good shape every year. Then, most likely, a planning period of 20-40 years should be used. These arguments clearly demonstrate that roads are an integer part of the whole forest management system. The mutual interdependency between forest management and roading shall be observed by all staff members.

It can be mentioned that mathematical models have been worked out which simultaneously optimize spacing and standard (Larsson, 1959, Larsson and Rydstern, 1968). These models contradict the conclusions advocated in this text in one aspect: road spacing can be substituted by road standard. However, estimates according to both methods give results which do not differ much. In any case, the mathematics of such simultaneous optimization is much more complex and tend to obscure the main purpose of this text - to demonstrate the structure of logging and roading problems and how they can be addressed with analytic tools.

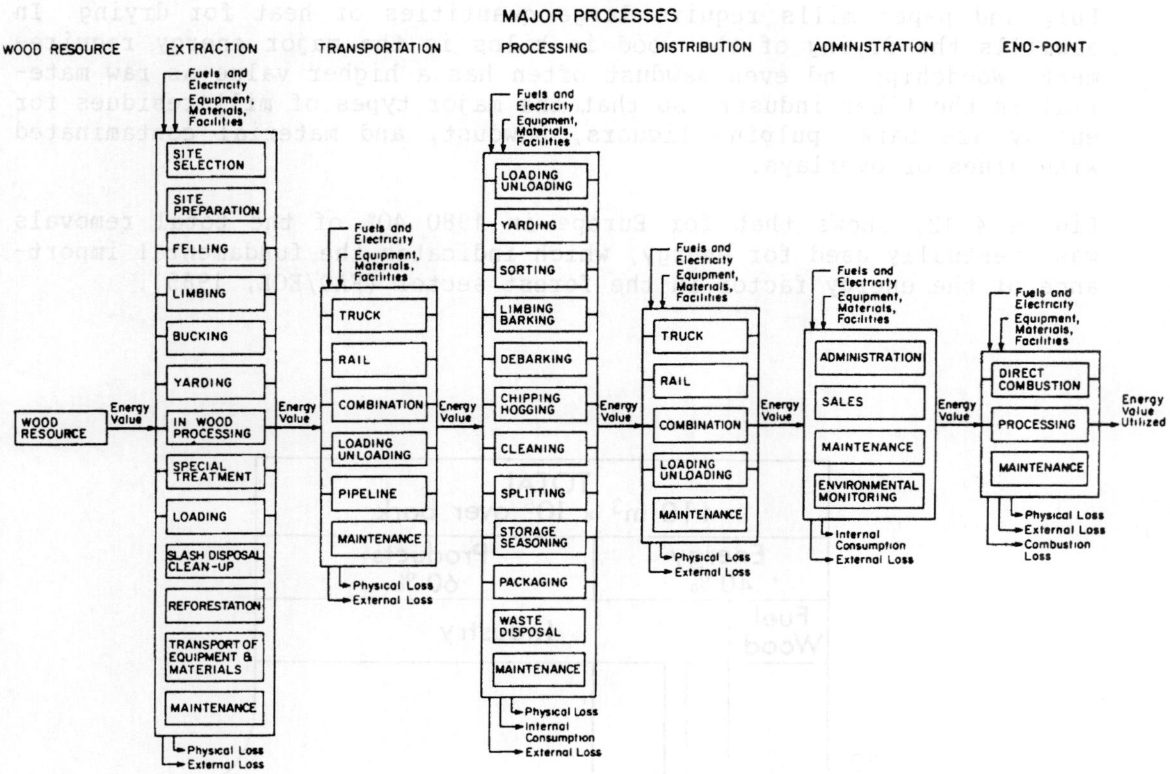

Figur 4.11. A generalized fuelwood energy trajectory. All or a portion of the subprocesses shown may be present in an application depending on the specific circumstances. (Baltic et al 1983)

Energy plantations of fast growing tree species in short rotation forestry is presently in a state of rapid development. Only the results of a study by Golob (1986) shall be quoted as indicative of the great potentials of energy or fibre production in such systems. Golob estimates the total cost of delivering wood chips from such plantations in Eastern Canada as around 31 Can\$ ODT^{-1} at a truck hauling distance of 30 km, employing highly mechanized methods for land preparation and other subsequent operations. About 30 percent of the cost is land preparation, about the same magnitude as the harvesting and transport cost (ODT = oven dry tonne, moisture content 8%, dry basis). The low price per tonne of oven dry wood chips delivered to pulp mills is quoted as Can$ 49,00 in 1985.

4.7 ENERGY IN SAWMILLING

Most forest industries are now very conscious of their energy costs. The energy content of the residues is, especially since the oil embargo of 1973 used increasingly to meet heat and power requirements in the mills. In Sweden, purchased fuels and electricity accounted in 1980 for 2,5% of the sales value of sawmills, 10,7% for pulp mills and 13,4% for the paper and paperboard industry.

Pulp and paper mills require large quantities of heat for drying. In sawmills the drying of the wood in kilns is the major energy requirement. Woodchips and even sawdust often has a higher value as raw material in the fiber industry so that the major types of mill residues for energy are bark, pulping liquors, sawdust, and material contaminated with glues or overlays.

Figure 4.12, shows that for Europe in 1980 40% of the total removals was eventually used for energy, which indicates the fundamental importance of the energy factor in the forest sector (FAO/ECE, 1985).

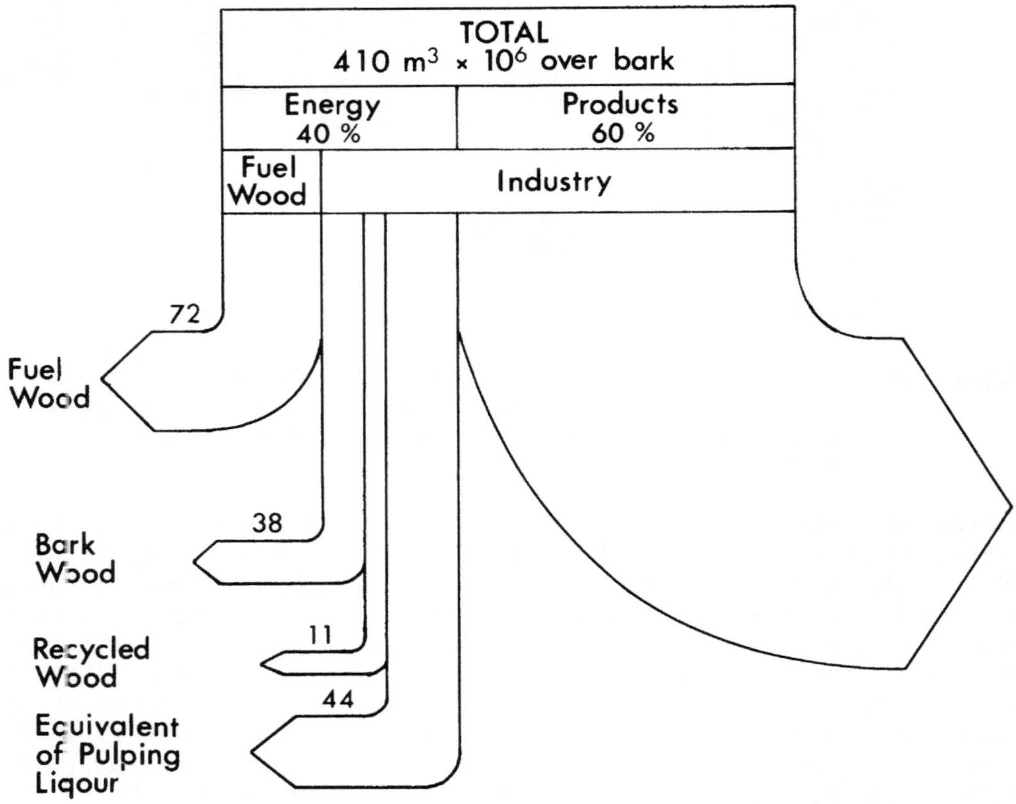

Figure 4.12. Total removals from the forests of Europe in 1980, by end uses.

Energy conservation in the pulp industry was particularly successful in the period 1975-85 and forecasts indicates that in some countries the pulp industry will be a net producer of energy. The energy systems in pulpmills are complex and will not be reviewed here. The energy consumption by sawmills for some countries is shown in table 4.9 (FAO/ECE, 1985).

Table 4.9 Energy consumption by the sawmilling sector in 1980.

Country	Electricity	Liquid	Wood	Other	Specific consumption GJ/m^3
USA	14%	3%	80%	3%	2,13
Finland	16%		84%		1,29
Sweden	16%	25%	59%	-	1,62

The energy use for a typical sawmill in Sweden is shown in figure 4.13 which shows that 85% of the total energy requirement was used for the drying of lumber in kilns.

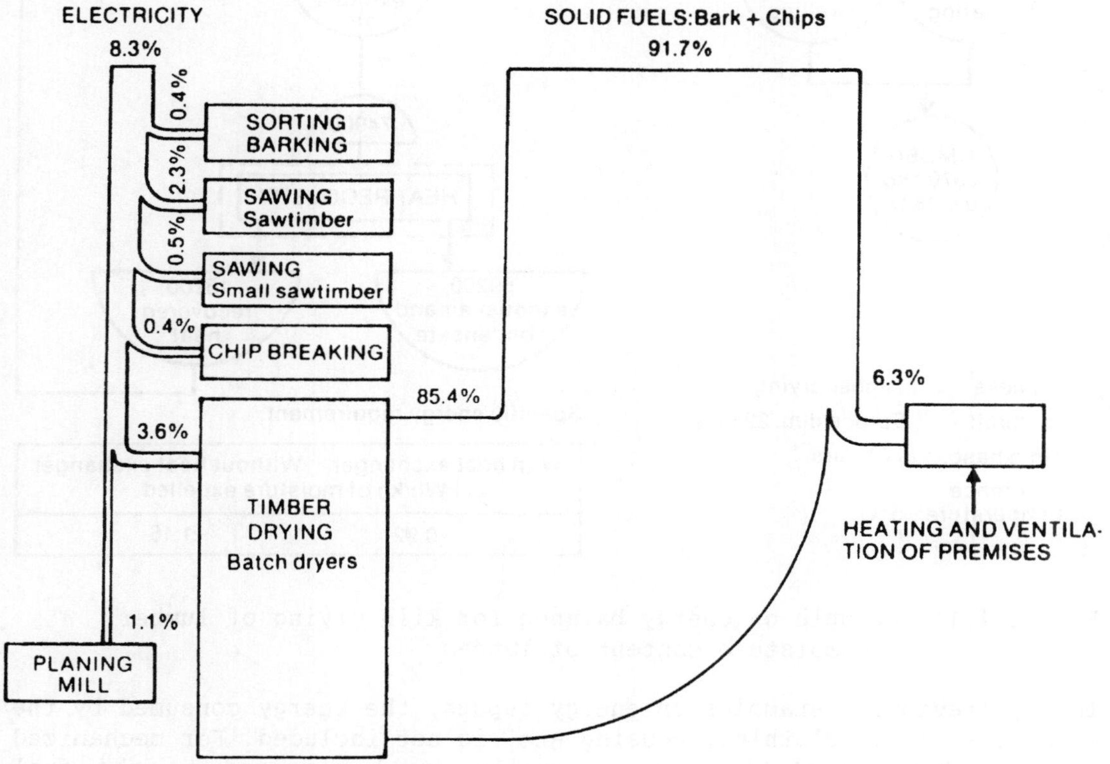

Figure 4.13. Example on the energy consumption in a Swedish sawmill. The plant handles approx. 20 000 m^3 of timber annually. The total energy consumption is 10,7 GWh year^{-1}. (UN/ECE 1982)

The energy balance for the kiln-drying process in a typical sawmill is shown in figure 4.14. In this case, every kilogram of moisture expelled from the kiln, required about one kWh.

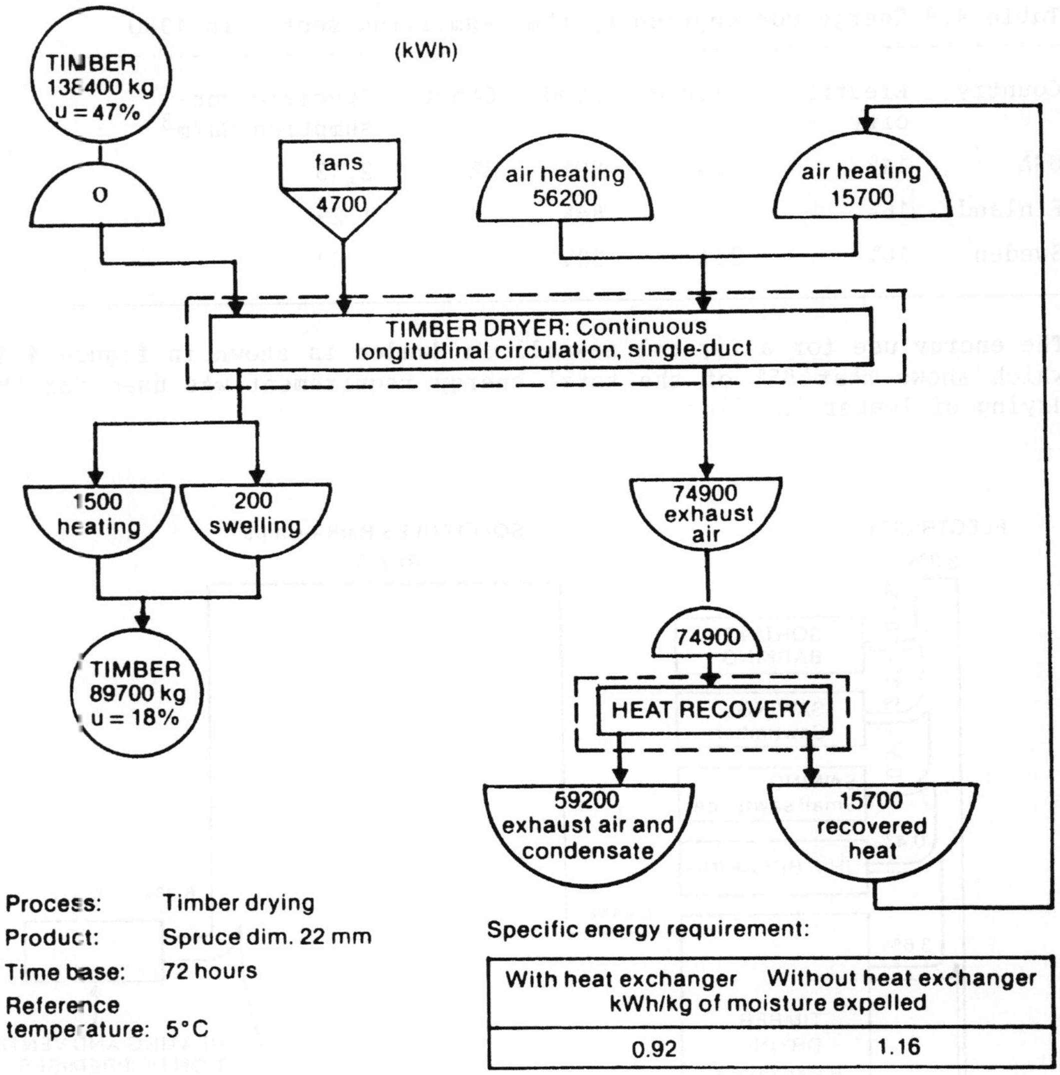

Figure 4.14. Example on energy balance for kiln drying of lumber.
u = moisture content of lumber

In the previously examples on energy inputs, the energy consumed by the workers as food, clothing, housing etc. is not included. For mechanized systems these quantities are generally small compared to the fuel inputs. It is also an arbitrary decision how much of a worker's total energy needs that should be accounted on the work. If all energy used up in his household and for the public services he and his family enjoys, is included, the energy cost of labour in a modern society is considerable. (See further Costanza, 1980)

A person has approximately the same ratio of efficiency as a machine in the conversion of chemical energy, from food or petroleum products respectively into work. Put on a bicycle ergometer the energetic efficiency is a maximum of about 25%, but in practical work it goes down to around 5% and often less.

The developments in agriculture during the history of mankind can briefly be characterized by increasing inputs of energy. The hunter in prehistoric times had an EROI of 1 to 10 where as for modern agriculture this ratio is the reverse, 10 to 1. This means that now an input of 10 calories is required to produce and deliver one calory of edible food. But the use of fossil fuels enormously expands the resource base that allow people to exploit natural systems. (For further details see e.g. Leach 1976 or Hall et al 1986.)

4.8 INPUT OF ENERGY AND LABOUR

4.8.1 Production Functions. The rapid mechanization of forest work operations during the last decades involves the substitution of human and animal labour for machines. Before this mechanization there were very few options to perform the work. The choice of technology or work system was as a rule a rather simple problem. Now, for each operation many possibilities exist and the technological development still goes on exploiting the general advancement in techniques and civil engineering. This creates needs for better analytic tools, through which a wise choice of technology can be made and a continuous adaptation to technical developments can be put in perspective. Much capital is now invested in the man-machine systems which must yield satisfactory profitability.

Let us assume that for a specific operation a number of man-machine systems are available and that figures on the production rate in each one of them are available or can be estimated. We than compute:

input of labour per unit of production = L
input of machines per unit of production = M
and plot the numeraires of L and M in a graph, the observations 1-6 in figure 4.15.

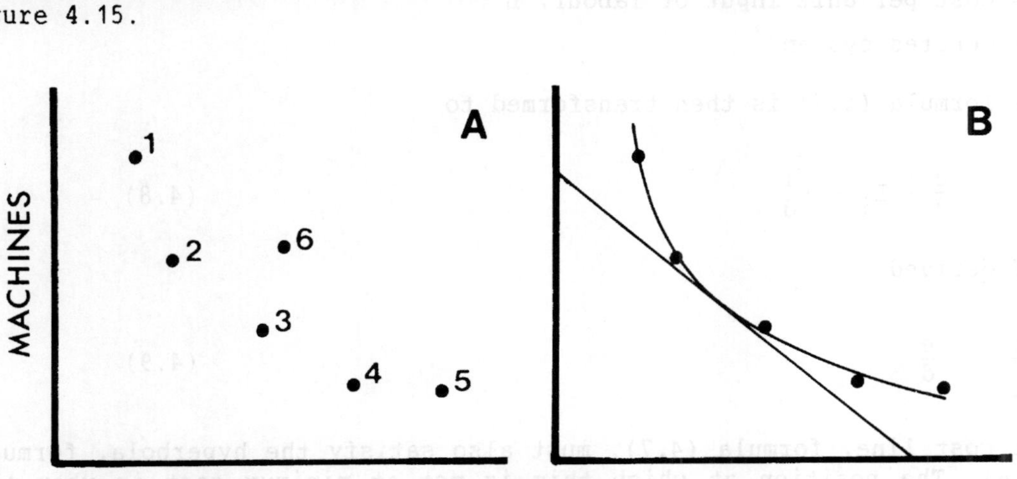

Figure 4.15. Production function (isoquant model) for deriving optimal level of mechanization. Graph A indicating observations of labour (L) and machine (M) inputs per unit produce in used systems. In graph B the observations are used to establish the isoquant (I). A cost line (C) is introduced. Opt = the input mix resulting in the lowest possible cost. (Sundberg 1979)

We can now observe that system 6 is not a good choice as it requires more input of labour than system 3 and at the same time more input of machines than system 2. On that basis we can rather safely discard system 6 from our list as an inferior and obsolete one.

The observations 1-5 in the graph are discrete points but in spite of this we connect them with a curve, figure 4.15 B, which has the form of a hyperbola. This is acceptable as it most likely should be possible to design systems located on or close to the curve at any place between 1 and 5. The hyperbola-like curve is then formalized in the equation:

$$M = a + \frac{b}{L} \tag{4.6}$$

a and b are constants which we can quantify on the basis of the observations 1 to 5.

This equation tells us the mix of labour and machines that are required to produce one unit, say one cubic meter of wood on different levels of mechanization.

In order to find the best system with the lowest cost per unit of production, say \$ per m^3, we introduce a cost line

$$C_i = d \cdot M_i + e \cdot L_i \tag{4.7}$$

C = cost per unit of production

d = cost per unit input of machines, M

e = cost per unit input of labour, L

i = denotes system

The formula (4.7) is then transformed to

$$M_i = - \frac{e}{d} \cdot L_i + \frac{C_i}{d} \tag{4.8}$$

and derived

$$\frac{dM_i}{dL_i} = - \frac{e}{d} \tag{4.9}$$

The cost line, formula (4.7), must also satisfy the hyperbola, formula (4.6). The position at which this is met at minimum cost is when the cost line is tangential to the hyperbola. We derive the hyperbola, formula (4.6).

$$\frac{dM}{dL} = - \frac{b}{L^2} \text{ and put } - \frac{b}{L^2} = - \frac{e}{d} \text{ and solve}$$

$$L = \frac{b \cdot d}{e} \qquad (4.10)$$

$$M = a + \frac{b \cdot e}{d} \qquad (4.11)$$

This is the mix of labour and machine givning the lowest cost per unit of production.

The model described above is often used in economics, especially for studies on the microlevel within the branch of the science of economics often termed production economics. We shall now give the model a general form. (See Sundberg and Svanqvist, 1987) The most convenient technique of representing several technical options with different mix of labour (v_1) and machines (v_2) is the isoquant

$$\bar{q} = f(v_1, v_2) \qquad (4.12)$$

When, as in most real cases, the rule of diminishing returns for increasing inputs of either v_1 or v_2 applies, the isoquant has a hyperbolic form and its slope is termed the marginal rate of technical substitution, since it measures the ratio at which the factors can be substituted against each other. It is assumed that the isoquant is based on the best technology for each available input mix (Cf. the rejection of system 6 in figure 4.15 A). As long as the change in production level does not effect the unit cost of v_1 or v_2 in the function, different production levels can be represented by parallel curves. (In figure 4.15 and in the formulae a production level of 1 m^3 is used but any level can arbitrarily be chosen.) Again we introduce the cost line

$$C = P_1 \cdot v_1 + P_2 \cdot v_2 \qquad (4.13)$$

where

C = total cost of producing one unit

P_1 = price of one unit of v_1

P_2 = price of one unit of v_2

We derive the two functions (4.12) and (4.13) and obtain

$$dq = f_1 dv_1 + f_2 dv_2 \text{ transformed to } \frac{dv_1}{dv_2} = -\frac{f_2}{f_1} \quad (4.14)$$

as dq = 0 and

$$\frac{dv_1}{dv_2} = -\frac{P_2}{P_1} \quad (4.15)$$

It follows that the tangential point between the isoquant and the cost line for lowest possible cost is determined by the equality of the two first derivates (4.14) and (4.15). This is the common point of the iso-cost line and the isoquant, closest to origin which therefore gives the lowest possible cost.

$$-\frac{f_2}{f_1} = -\frac{P_2}{P_1} \text{ or } \frac{P_2}{P_1} = \frac{f_2}{f_1} \quad (4.16)$$

This analytic model of input/output relations was first developed by Cobb and Douglas (1928) in macro studies on the national level. They derived the following formula based on statistics from the US economy.

$$Q = A \cdot K^{\alpha} \cdot L^{1-\alpha} \quad (4.17)$$

where

Q = Gross national product

K = Capital stock

L = Labour force

A = Constant

This function seems to be adequate only as long as the labour share is constant. α then usually takes a value of 0,25 and thus $1-\alpha = 0,75$.

Many economists have later tested this model with varying success and often tried to introduce other input variables in attempts to achieve a better fit. Energy is one such variable introduced by many followers to Cobb-Douglas.

4.8.2 Application. We now have a rather simple model for establishing the optimal level of mechanization. But some rigid provision must be met in using it. The end products of the work systems in figure 4.15 the systems 1 to 5 must be identical, e.g. for logging operations the processed wood must be in the same physical form and delivered at the same point. Similarily, the work conditions of the systems must be identical, e.g. identical trees and stands, equal extraction distances etc. Such provisions for the data base are seldom at hand. However, on the basis of time studies or log book data the production rates can be recalculated to "iso" cases (iso = equal). Our knowledge of the influence on the production rate of such parameters as size and form of the trees, terrain variables, extraction distances etc are now so good that the rates for all systems can be recalculated to one particular iso-case.

The input variables should be quantified in some physical scale, not money, and a unit cost of this inputs must be possible to establish. For labour the input is best quantified as time, e.g. mandays or man-hours. If the labour employed in the systems have different wage rates, two possibilities are open. One is to use an average rate for the work team. The second is to use the wage rate of the individual workers located around the expected optimal system. Possibly both numeraires can be used in alternate calculations in order to explore if it has any influence on the choice. Such an exercise can be part of a sensitivity analysis, which always should be made to elucidate the impact of changing ratios of the cost of labour and machines. A choice of technology is always binding for the management for some years ahead as new techniques cannot be introduced overnight and always carry considerable expenses for introduction and training. The analytic model does not give the final answer, only some basic indications to be put in the context of forward planning of the operation.

In general economics, the input of machines is often quantified as capital, as money. Several problems are associated to this measure. First, money is not a physical input and it leads to the rather strange notion of putting a unit cost on money. This is possibly acceptable in such trade as banking but hardly so in engineering studies. Secondly, the monetary value of machines and installations is always subject to considerable uncertainties. Many researchers prefer to use the insurance value of these assets. But insurance values on assets may often be determined using other criteria than production economics. For machines several other scales have been suggested as weight, number of moving parts etc. Koroleff (1961) aggregated the HP-rating of all the machines in a system as a measure of the level of mechanization and on machine capital. A third scale was suggested by Sundberg (1979, 1982). The fuel consumption of the machines, which can be looked upon as one step further to Koroleff's approach.

The underlying assumption for using fuel consumption as a parameter for capital input is that fuel (or energy) is a direct input for reducing the input of human or animal work and it can at the same time represent the cost of using the capital in the form of machines. The value of a machine is decreasing with the aggregated input of energy over time and the rate of the declining value should logically be proportional to the fuel input. Put in other words, energy for running the machine wears it out and is also the major cause for repairs and maintenance. Therefore, we shall focus our attention on the validity of the hypothesis that the fuel consumption of a machine is a good indicator of the total cost for using not only this specific machine but of the cost of all types of forestry machines.

4.8.3 Studies on the Cost of Machine Use. In a study published 1979 Sundberg compiled machine log book data obtained from a number of Swedish forest enterprises. The data were from the year 1977 and represent machines of various makes and of several designs with regard to production functions in logging and primary transport. The result of the analysis shows a close similarity in aggregate costs, excluding personnel costs, expressed per unit fuel consumed,not only within brand and functional groups but also among different types of machines, varying from power saws to processors.

In 1982 Sundberg analysed machine log book data from three large forest enterprises in Sweden in order to verify the hypothesis that the aggregate cost of forest machines, excluding labour costs, is the same, expressed as costs per unit fuel consumed, for all machine categories. The data were from 1979, representing 913 individual machines of eleven functional types and a large number of brands and combinations of processing components.

The average costs derived from the material included in the 1982 study support the hypothesis of equal aggregate cost per unit fuel consumed, though the material contains large variations among machines of the same brand and the same functional type. A comparison of cost data contained in reports from Canada (Bjerkelund, 1979), USA (Cubbage, 1981) and Hungary (Rumpf, 1986) gives further support to the hypothesis.

Based on indications derived from the analysis of the 1979 machine log book data Sundberg suggests that the hypothesis of equal aggregate cost per unit fuel consumed needs to be qualified or modified to incorporate considerations about intensity of use. Capital costs, and probably also repair and maintenance costs, are affected by the number of days or hours a machine is in use per year. Consequently, different yearly utilization will cause different machine costs. He also observed that the technical maturity of machines may cause variations in costs, so that a machine which has had the benefit of experience from a longer period of use has a lower aggregate cost than a newly introduced machine, and that machines which have been available on the market from some time and are manufactured in large series for use also in other fields of work than forestry have still lower costs. Based on these observations Sundberg proposes a matrix, as presented in table 4.10, for adjusting the uniform machine costs, expressed per unit fuel consumed, to practical production situations.

The machine costs in the tables 4.10 and 4.11 are expressed as ratios, 100 being applicable for the most commonly used machines during a normal intensity of yearly use. The cost level in absolute terms (e.g. US$ per litre fuel) of the ratio 100 will change with time because of inflation and other changes in the economy. It is also specific for each country or location. A procedure to establish this cost level is suggested later on in the text.

Table 4.10 Matrix for the adjustment of an uniform machine rate per unit of fuel consumed to practical situations

Machine Characteristics	Intensity of machine use Low	Normal	High
First generation machines	125 or >	115	105 or <
Special forest machines, technically "mature"	110-115	100	90-95
Machine manufactured in large numbers, also for other jobs than forestry	100 or >	90	85 or <

In 1985 Svanqvist presented an analysis, based on a large quantity of material on machine log book data and covering the period 1971-1983, in which he verifies Sundberg's hypothesis of equal costs of machines and attempts to establish the cost relations of Sundberg's adjustment matrix, table 4.11. He also analyses the trends in costs and the change in relative importance of the major cost components.

Table 4.11 Matrix (in table 4.10) as modified by Svanqvist (1985)

Machine Characteristics	Intensity of use Low	Normal (average)	High	Very high
First generation (less than 5 years on the market)	135 or >	115-120	110 or <	
Mature (more than 5 years on the market)	120-130	100	80-85	
High-way trucks				60

The hypothesis of equal aggregate cost, expressed as cost per unit fuel consumed is verified again by the material analysed by Svanqvist. He also concluded that the effects of intensity of use and technical maturity can be expressed by a matrix, as Sundberg suggested. The classification of machines in maturity classes is, however, disturbed by the fact that most machines have attachments which themselves are of varying technical maturity, different from the maturity of the basic machine unit, and that repair and maintenance costs, in most cases, can not be assigned to the various components of an integrated machine assembly. In the statistical material used by Svanqvist only that for high-way trucks is clearly to be referred to the category "machines manufactured in large series also for use in other fields than forestry".

The precision of using fuel consumption as an indicator on the input of capital into forest operations and also as a determinant of machine cost is hampered by he scantiness of the data base. Few enterprises up to now kept detailed records on fuel consumption and machine costing procedures are also inconsistent between enterprise. The numeraires (ratios) in the matrix, derived from a rather scanty data base, should therefore be regarded as tentative.

A definite advantage of using fuel (energy) is that all machine units in a system can be aggregated into one measure. In production studies it is always highly desirable to aggregate the inputs into as few variables as possible, otherwise the models become to complex and results difficult to interpret and use. The problem of aggregation is often the key problem in production economics and in the design and quantification of production models. It seems that the use of fuel as a measure of both the input of energy and of capital could contribute to mastering the problem of aggregation. Capital assets, and also materials, always represents energy used up in their creation, generally termed embodied energy. This embodied energy, or a major part of it, is consumed while using and wearing out the machine. Only the embodied energy in materials being re-used e.g. scrap iron, is recoverable where as the balance finishes as waste matter or waste heat in the energy system and also as a zero value in the economic process. As the fuel input as aggregated over time of use is parallel to the wearing out of a machine, there is considerable logic of letting fuel represent also input of capital. There are also technical reasons. The higher the input of energy, the faster is a machine worn out and the more repairs are required. Take as an example a passenger car driven at 70 km/h and 100 km/h respectively. Its cost both per unit of time and unit length of travel is higher for the higher speed as also is the fuel consumption. Such correlations are used to minimize travel cost. The proposal here is to use such corelations in a wider system context.

The numeraires in the matrix (table 4.10 and 4.11) are ratios, 100 corresponding to a specific cost per unit fuel (SEK = Swedish Crown, exchange rate used, 1986 1 US$ = 7,50 SEK):

in Sundberg's study, 1979	100 = 15,10 SEK	2,01 US$/liter diesel oil
in Svanqvist's study, 1984	100 = 20,40 SEK	2,72 US$/liter diesel oil

(Note: The actual exchange rates were approx.: in 1979 1US$ = 5,50 SEK, in 1984 1 US$ = 9,00-9,80 SEK and in 1987 1 US$ = 6,30 SEK.)

The cost increase 1979 to 1984 was 35% which corresponds closely to the rate of inflation in Sweden.

The machine cost per unit fuel consumed is time and country specific as is the cost of labour. It can be derived for each locality from cost records and log book data from enterprises operating in the area. However, in many instances a researcher or planner is faced with a situation where records on the use of forest machines are not available. This is often the case in projects in developing countries. Then machine cost can be assessed with traditional costing methods as outlined in the chapter 1 and the Appendix B to Chapter 1. Such estimates can, however, be supported by parallel cost estimates using fuel consumption as cost indicator along the following line.

1. The fuel consumption of the different forest machines can be estab lished using data from operations elsewhere in similar setting. The fuel consumption can also indirectly be assessed using the power rating of each machine and the Work Load Factor (WLF) in using the machine for the actual operation as described in the Appendix C to Chapter 4.

   Fuel consumption = kW-rating · fuel consumption per kWh · WLF

2. Then, an unit cost per liter fuel shall be derived but when no forestry machine cost data are available, this must then be done in an indirect way. In each country or each setting there are always in operation some similar machines to forestry ones, e.g. crawler tractors. Cost data and fuel consumption for these machines are then collected and the total machine cost per liter fuel computed. The cost for each machine type is then entered in the matrix (table 4.10 and 4.11) whereby the cost per liter fuel in each box is determined. The forestry machines are also classified according to the matrix and the cost per liter fuel as derived for each "box" in the matrix is then applied.

4.9 BIOPHYSICAL ECONOMICS

Economics is the science which deals with allocating scarce resources among competing means to satisfy the wants of mankind. The results of economic research have no doubt had an immense impact on material developments. The views of economic experts have guided governments, firms and individuals in the management of their affairs. However, economic science is established as a discipline with its own jargon and a multitude of analytic tools to elucidate the allocation process and to forecast future events. It is interesting to note that many of the great peers in economics in the past had no formal training in economics as we know it today. Now, the economists seem to form a quite coherent group with their own language and epistomology. They are not particularly keen on listening to outsiders, including the authors of this book, or to admit them into their system.

Fundamental to economics is "the economic man" and the "market place" even when allowance is made by economists for imperfections in these concepts. Land and labour are the basic production factors, jointly creating a third one, capital. The scale in economics is money.

But economics is not the only way, and perhaps not the best way, to examine the flow of goods and services in an economy. For example, researchers in biological sciences some decades ago more systematically began to describe and quantify the biological growth processes and the behavior of ecosystems as a flow of energy and matters, based on the laws of thermodynamics. Some of the pioneers were the two brothers H.T. and E.P. Odum (1976). Even such topics as pollution was described as a phenomena of energy.

One of the first professional economists to enter the road towards what later by Cleveland et al (1984) was termed biophysical economics was the Romanian born but North-American researcher Nicholas Georgescu-Roegan (1971, 1975, 1979). Some of his critical views on traditional economics, of particular interest for forestry were:

(i) Traditional economics is not relevant to agriculture. It does not take into account that in many agrarian economies a large part of the population often is idle and superfluous to the available land and capital.

(ii) Traditional economics are adapted to deal with line production in factories in which idleness of all agents can be eliminated. Agriculture produces in series.

(iii) Traditional economics regards only one generation, ours, as future ones cannot bid on the present day market.

He then crystallized his thoughts in describing the economic process as the entropic transformation of valuable natural resources (of low entropy) into valueless waste (of high entropy).

In production economics, until comparatively recently, the role of energy was not considered in much detail or even at all. Energy was cheap and up to 1973 became steadily cheaper. Mechanization was viewed merely in terms of cost of capital for machines replacing men. The oil embargo in mid 1973 with its sudden jump of energy prices, followed by a similar rise at the end of the 70's upset the economics on all levels throughout the world and seems to have disrupted the validity of traditional economics. A new breed of researchers came out using energy instead of money as the key parameter in exploring and modelling production processes. They challenged traditional economics for not being able to deal with the change. Drucker (1982) stated that "both as economic theory and as economic policy Keynesian economics is in disarray" and Leontief (1982) described many economic models as unable "to advance in any perceptible way, a systematic understanding of the structure and the operations of a real economic system".

Although much of economic analysis is based on very simple relations assessed with great analytic complexity, considerable critisism has also been made of the complexities of those fewer economic models that use statistical analysis of very large data bases. Georgescu-Roegan suggested "to pick up a small number of relevant variables, instead of relying upon the computer to juggle with scores of variables and thus losing all mental contact with the dialectic nature of economic phenomena". Some of the energy researchers go so far as to suggest that energy analysis is the all embracing science of economics and that traditional economics is only one of its special sub-sciences. Underlying this argument is that the law of thermodynamics is or should be governing all human activities and that the entropy law is the taproot of scarcity. Lately Cleveland (et al, 1984) took a middle way, coined the term "Biophysical Economics", and stated the following hypothesis:

"We approach macroeconomics from a thermodynamic perspective that emphasizes the production of goods, rather than the neo-classical perspective that emphasizes the exchange of goods according to subjective human preferences. Production is the economic process that upgrades the organizational state of matters into lower entropy goods and services. Those commodities are allocated according to human wants, needs and ability to pay. Upgrading matter during the production process involves an unidirectional, one-time through put of low entropy fuel that is eventually lost (for economic purposes) as waste heat. Production is explicitly a work process during which materials are concentrated, refined, and otherwise transformed. Like any work process, production uses and depends on the availibility of free energy. The laws of energy and matter control the availability, rate, and efficiency of energy and matter use in the economy and therefore are essential to a comprehensive and accurate analysis of economic production."

They do not disqualify standard economic models but suggests that resource allocation can be understood more clearly by explicitly accounting for the physical constraints imposed on economic production.

On the micro-level, Slesser (1978) in particular applies the same principles in analysing production and agriculture has been studied from an energy perspective by many researchers such as Leach (1976), D.O Hall (et al 1985) and many others.

Further support for the thermodynamic perspective on the production process is the investigation by Costanza and Herendeen (1984) in their input-output (I-O) analysis of 87 sectors of the U.S. economy for the years 1963, 1967 and 1972. They showed that the economic output in all these sectors was highly correlated to the energy input, if households and government were included as endogenous sectors in the model. They concluded that the embodied energy, calculated in the comprehensive way they suggest, is a good predictor of economic value and that the economy operates on an energy theory of value as the value is proportional to an appropriate energy indicator. (Note: the output of households and government are labour and government services. In economicstudies they are as a rule considered as exogenous sectors having a controlling position outside the economy. Costanza and Herendeen suggest that the intended use of the analysis is decisive.)

This approach to the production process is developed further by Hall et al (1986). They illustrate the two schools of thoughts in two graphs and use the expression "The Ecology of the Economic Process" for the thermodynamic perspective on production. They admit that the theory of this concept is not fully developed and that both models suffer from imperfections. This ongoing discussion on the basic paradigms of economics is reviewed as it is likely to have considerable relevance on forest resource economics.

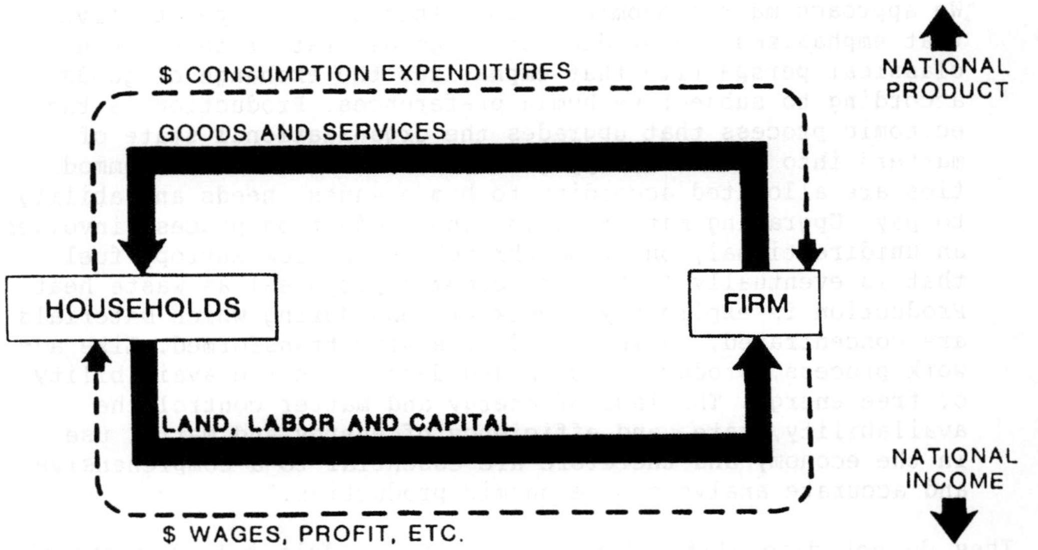

Figure 4.16. The neoclassical flow model of economic production. Households sell or rent land, natural resources, labour and capital to firms in return for rent, wages and profit factor payments). Firms combine the factors of production and produce goods and services in return for consumption expenditures, investment, government expenditures, and net exports. (From Hall et al 1986)

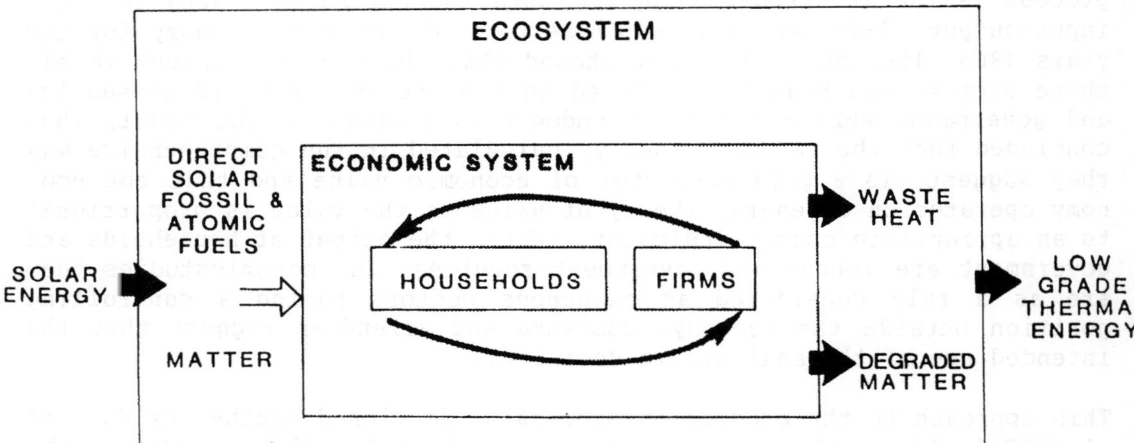

Figure 4.17. Economic production from an energy perspecitve. A continuous input of high-quality, low-entropy fuels enters the economic system, and the economy uses these fuels to upgrade natural resources, driving the circular flow between households and firms in the process. The fuel is degraded and put back in the environment as low-quality, high-entropy heat. (From Hall et al 1986)

Forestry research along the lines of "Biophysical Economics" up to now has been comparatively scarce, in particular in the field of production techniques and operations. As forestry in the long term is a problem of energy transformation, from the sun to trees to whatever goods and services are aimed for, thermodynamic analyses hould give additional insights into the mechanisms of forestry and provide a safer ground for decisions and plans both in the short and long term. In particular it seems that such energy analyses will be increasingly important if petroleum prices continue to escalate, as they must eventually due to the finite nature of this energy source. The process analysis of forest work operations based on energy use and the suggested linkage between machine cost and fuel consumption is an application of biophysical economics on the micro level in forestry. Up to now, the data base hasbeen too scanty and too inconsistent to allow a more firm appraisal of its validity and usefulness. But experiences from other sectors of production strongly support that such studies are meaningful and valuable to advance forestry developments. The aspiration is to improve our battery of research tools to come to grips with reality and facts with main bearings on the future of forestry. A relevant question is whether energy analyses, being of necessity circumscribed by the laws of physics, is ultimately of greater utility in predicting and analyzing economic activity than economic analysis which has no such constraints.

4.10 CRITICISM OF ENERGY ANALYSIS - THE CONVERGENCE ISSUE

Net energy analysis requires that there be a point beyond which sufficiently remote, indirect energy flows do not significantly affect the estimate - a point called convergence. Only if this assumption holds will the result be realistic. In principle, one is faced with interminable regresses into the past and future, associated with the energy inputs to and outputs from the system under study. Energy analysis would be unmanageable unless distant energy flows can be correctly aggregated in normative figures, safely ignored or "truncated".

No one has as yet presented a methodology for establishing convergence boundaries. The criticism is that a large number of truncated flows, each insignificant, may add up to significant quantities.

This scientific shortcoming of energy analysis has largely discouraged energy analysis and the credibility of such research. However, the criticism should have little relevance for studies of the type exemplified in this text, process analysis of rather small systems with limited extension in time. It can also be argued that other studies of the economics of biomass production, e.g. using money as scale, embodies "truncations" of physical flows of assets, energy or materials. It seems that traditional economic theories can be questionned on the same grounds.

Those who want to know more of the convergence issue could use Reaven (1986) as an introduction.

APPENDICES TO THE CHAPTER 1. PROBLEM IDENTIFICATION

A. NOTES ON THE ERGONOMICS OF FOREST WORK
1. Physical Work Capacity
2. Nutrition and Work Capacity
3. Heat Stress
4. High Altitudes
5. Social and Psychological Factors
6. Work Motivation
7. Cultural Traditions
8. Health

B. DERIVATION OF MACHINE COST
1. Depreciation
2. Interest
3. Insurance
4. Taxes
5. Operating Labour
6. Fuel
7. Oil and Grease
8. Servicing and Repairs (Except Tires for Hauling Rigs)
9. Tires (for Hauling Units Only)

C. NOTE ON ACCOUNTING PRICES OR SHADOW WAGE RATES FOR LABOUR AND LAND IN FORESTRY DEVELOPMENTS
1. Labour
2. Land
3. Conclusions

APPENDIX TO THE CHAPTER 1. PROBLEM IDENTIFICATION

A. NOTES ON THE ERGONOMICS OF FOREST WORK[1)]

1. Physical Work Capacity

Logging operations are heavy, often so heavy that the physical work capacity of the workers puts a practical limit on the work output. The following factors relating to the individual should be considered in the first place: age and sex, body size, physical fitness, nutrition and health. The environment, especially heat and high altitudes, may also impede work capacity.

In developed countries physical work capacity reaches its highest level by about 20 years of age. At about 35 years, it declines rather rapidly, to a level of about 60 percent at 60 years of age. This decline, which for various reasons may proceed faster in developing countries, can partly be compensated for by skill and experience.

It has been observed that forest workers in developing countries are rarely from the older groups. Whether this derives from the individuals' own evaluation of the work as being too heavy for him, or from the employer's selective recruitment of young workers, it is an indication that forest work may often be extremely taxing in relation to the physical work capacity. Experience from developed countries shows that it is not only the total workload over a shift that creates a reluctance toward heavy work, but peak workloads of rather short duration also make the job unattractive.

As regards body size, there is some evidence that the physical work capacity is closely related to the average weight of a population of forest workers, although relatively big deviations in individuals may occur.

Anthropometric data on rural labour from developing countries seems to be scarce. The following figures on the weight of adults, taken from Clark and Haswell (1970), give some indication of the variations in body weight of rural workers. (See table 1.1, p. 156)

Physical work capacity of a person is often expressed as his maximal ability to take up oxygen in his circulatory system, e.g. maximal capacity of oxygen intake. This can be assessed in tests on a bicycle ergometer. Setting this figure in relation to his body weight (e.g. millilitre oxygen per kg weight) indicates his fitness for work and sports which require much movements, for example felling, running and skiing. See further Åstrand and Rodahl, 1970.

---

1) This appendix draws to a large extent on Sundberg, U: Harvesting man-made forests in developing countries. A manual on techniques, roads, production and costs. FAO, Rome 1976.

Table 1.1. Examples on body weights of rural workers

	Weight in kg Males	Females
1. India, a village near Mysore coast	48-51	40-50
2. Eastern Asia	50-55	
3. Burma, primitive agricultural community	50	43
4. Philippines, primitive pagan cultivators	48	
5. Sudan	57	
6. Uganda	58-64	51-59
7. African pygmies	40	
8. Latin America	56	
9. Europeans	65	

For further information on work capacity see Åstrand and Rodahl, 1970

The body size of females is smaller than that of men. In developed countries, females weigh about 25 percent less than men and their physical work capacity is correspondingly lower. There are indications that the difference in weight between men and women in developing countries is less accentuated, in the order of 15 percent.

2. Nutrition and Work Capacity

It is well known that the level of nutrition as well as the balance of the diet is unsatisfactory in many developing countries. One of the most prominent symptoms of prolonged lack of food is the obvious loss of work capacity in its broadest sense. There is an appearance of lethargy and sluggishness; movepauses; any continuous effort is avoided as far as possible. Superficial observers have taken these conditions, commonly found in poor people living in hot climates, for laziness; sometimes it has been called an ethnical characteristic, or an enviable philosophical attitude toward life. When those explanations were proposed, no account evidently was taken of important social factors like undernutrition or malnutrition.

Undernutrition or malnutrition does not affect only physical work capacity. The psychological functions are also impaired. Although the following quotation, from an FAO study on "Nutrition and Working Efficiency", is based on experimental tests of relatively severe starvation, it throws some light on reactions and attitudes often encountered in the employment of an undernourished labour force.

"The most obvious mental changes in cases of starvation are emotional depression, and apathy and lassitude of the afflicted persons. Conversation is often restricted to a minimum; there is even a certain hostility toward others. Moodiness and irritability are other features regularly encountered and are in part due to a sense of incompetence. The depression is coupled with a general lack of drive and of initiative. Although some observers have reported a deterioration of mental capacity, actual impairment of intelligence has never been proven. The intellectual powers are unaffected but are simply not applied. As initiative and ambition are subdued, interests are narrower and it is found impossible to concentrate for long on any task."

"The general picture is often that of plain stupidity. Although most of the more detailed evidence has been obtained in relating severe starvation experiments, and although there are very litle precise data from less severe but chronic undernutrition, it appears justified to apply the acquired knowledge equally to those chronic states, especially as the general picture of apathy and lack of initiative is a common observation in many undernourished populations. Dietary measures have, in most cases, been successful in re-establishing normal personality patterns, but single traits of the changed personality may persist for a long time. The implications of the difficulties that are encountered in developing countries when untrained labour is employed may well be caused by chronic undernutrition and might be relieved by dietary improvements."

Undernutrition is not uncommon even in areas with an abundance of easily found edible items like vegetables, fruit and fish. The reasons are either lack of knowledge about basic nutritional needs or various taboos related to certain food items. These taboos prevent proper eating, for example, during certain periods when a fast is observed, or they are directed toward certain groups or are generally applied on certain food items. Rules concerning food are very difficult to change, but experience has shown that success can be achieved if a nutritional programme is planned within the existing framework of food restrictions.

The adverse effect upon worker efficiency of inadequate food consumption is well documented, but few observations from forest work are available. One quotation from a study in Jamaica (FAO Report TF/IND/112 SWE) is as follows:

"Studies carried out on plantation forest workers in Jamaica indicated that they did not receive enough calories for optimum work output. On the basis of a measured sample of workers, it was found that there was a total energy deficit per 24 hours of 1 500 calories per man. For the average man, the basal metabolism required 2 000 calories, walking to work a further 1 000 calories, leaving only 500 calories for forest work compared with an estimated need of 2 000 for a day's work."

"The Jamaican result indicated that either optimum work could be performed at the expense of body weight or optimum work was not performed. Plantation establishment costs on hilly terrain were US$ 250 per acre and it was felt that optimum work was not being performed and contributing to the high cost. Estimates were made on the effect of a supplementary on-the-job feeding programme and indicated that a gross saving in plantation costs for a 40 000-ha plantation programme could come to US$ 8 million for Caribbean Pine grown on a 20-year rotation for sawlogs."

It is evident that the level of nutrition plays a profound role on the work output but to what degree can be established only on the basis of measurement of the caloric requirements of the work in relation to the actual intake of food. It is also evident that lack of food contributes to lower resistance to diseases and leads to higher absenteeism and higher rates of accidents.

Widespread malnutrition and undernutrition cannot be eliminated other than as part or a general programme of economic development. However, there are certain measures by which employers and enterprises can ensure an adequate and balanced diet for their own employees. These measures consist of the establishment of non-profit-making food stores or of canteens at the working place. Apart from humanitarian considerations, the extra cost involved should be more than compensated for by increased production.

3. Heat Stress

Man has a low energy-converting eficiency and uses only about 25 percent of his total energy input. To cope with the heat produced by the other 75 percent, a rise in body temperature takes place (up to about 39° C, momentarily higher, during a long period of work of moderate to high intensity) and may be considered beneficial, because a high temperature facilitates the passage of heat to the surface of the body followed by the elimination of heat at the body surface (at rest having a temperature of about 34° C). As blood is the main carrier of heat from the working muscles to the body surface, an increased environmental temperature results in a need for more efficient circulation of the blood. A larger quantity of blood is required to transfer the same amount of heat when the temperature gradient decreases.

Thus, a hot environment causes strain to the circulatory system and heat transportation competes for the oxygen supply to the muscles, reducing the working capacity. Furthermore, as heat elimination at the body surface of a hard-working man is dependent mainly on evaporation, that is sweating, a high humidity and low air velocity will be particularly detrimental to the man's working capacity and the risk of (sometimes fatal) heat stroke increases considerably. However, working capacity under heat stress can be somewhat improved by acclimatization, the physiological adaptation which takes place particularly during the first week of work in hot conditions and results in an increased ability to eliminate heat by sweating and so on. (See further "Acclimatization" later in this appendix.)

Heat stress in forest work is elaborated in an FAO report by Axelson (1974). Anyone concerned with work operations in a hot and humid environment is advised to study this report from which a few salient conclusions are extracted in the following text.

Indicative tolerance limits to heat stress

To assess the limits of tolerance to heat stress is difficult and there is some inconsistency among the proposals on this matter in the literature published in past years.

The attempts made to establish these tolerance limits, usually based on so-called heat stress indices, are built up of such parameters as dry and wet bulb temperature, globe (radiant) temperature, air velocity and sometimes also work rate. As forest work is moderate to very heavy, in the hot, humid (⩾ 90 percent relative humidity) and calm environment of a forest in the tropical region, 27-29° C (dry bulb) indicates the range where the hazard of heat stress becomes considerable and heat strokes might readily occur if the workers are not highly acclimatized and if insufficient resting time is given for them to cool off.

The following Table 1:2 suggested by Axelson (1974) gives some guidelines to the level of reasonable tolerance limits which are greatly influenced by the work rate as well as the state of acclimatization. It must be strongly emphasized that the suggested limits do not necessarily protect all workers from heat stress disorders, although they are acceptable to the great majority of well-acclimatized workers.

Table 1.2. Limits of tolerance to heat stress in forest work
(A moderate to high state of acclimatization is assumed)

Activity	Energy expenditure	WBGT °C
Very heavy work	> 400 Kcal/h	27
Heavy work	300-400 Kcal/h	29
Moderate or light work	< 300 Kcal/h	31

WBGT = 0,7 t_w + 0,3 t_g, where t_w = aspirated (psychrometric) wet-bulb temperature and t_g = globe temperature. WBGT = index for temperature (T) based both on wet-bulb (WB) and globe (G) temperature. Wet-bulb = thermometer with bulb covered with wet cloth - it indicates lower temperature because it is cooled by evaporation which in turn depends upon the moisture present in the air.

Natural wet-bulb temperature is sometimes used (t_{nw}; measured by covering the bulb of an ordinary mercury-in-glass thermometer with a wetted wick and not shielding for radiation and prevailing wind).

WBGT = 0,7 t_{nw} + 0,2 t_g + 0,1 t_d is preferable, where t_d is the dry-bulb temperature.

The air velocity is assumed to be ⩾ 0,5 m/sec. At air velocities ⩾ 1,5 m/sec the limits might be increased by 0,5° C. WBGT should be used on the basis of a time weighted average.

The sweating rate is used as an index integrating heaviness of work and exposure to heat. Lundgren (see Åstrand and Rodahl, 1970) has enlarged the scheme, Table 1:13, p. 33, as follows:

Table 1.3. Grading physiological work load acc. to sweating rate

Work load	Very low	Low	Moderate	High	Very high	Extremely high
Sweating rate, l/h		< 0,2	0,2--0,4	0,4--0,6	0,6--0,8	>0,8

The World Health Organization (WHO) recommends a maximum of 4 l per shift.

Allowance in Time consumption or productivity reduction

Contrary to the industrial work environment, climatic conditions are not possible to manipulate. Therefore, heat stress in forestry means that a productivity reduction has to be accepted, as a lower work rate is necessary when heat stress increases or frequent breaks for rest and cooling off have to be inserted. However, work in the hot season can also to some extent be facilitated by starting early in the morning and avoiding it during the hottest hours of the day.

Only sparse data are available about the magnitude of productivity reduction due to heat stress, particularly as far as forestry is concerned. However, considering experiences from heat stress studies in the context of deep mining (that is, in a hot and humid work environment), the productivity reduction suggested by Axelson (1974) is indicated in Table 1:4 below (presuming a moderate to fairly high state of acclimatization of the labour force).

Table 1.4. Time consumption or reduction of productivity in heavy forest work due to heat stress (wind velocity assumed to be ≈ 0,5 m/sec)

Dry bulb temperature, °C	Relative humidity	Allowance in time consumption	Productivity reduction
26	90%	0%	0%
28	90%	10%	10%
29,5	90%	25%	20%
31,5	90%	55%	35%
33,5	90%	185%	65%
33	70%	45%	30%
35	70%	100%	50%
37	70%	550%	85%

The table shows that heavy work should preferably be performed in the early morning or the late afternoon and should be avoided during the hottest hours of the day. The workers should preferably be semi-nude or wear clothes which permit as free evaporation of sweat as possible. In this context it should also be recalled that working in the heat means loss of considerable amounts of water due to sweating.

Water and salt supplements

A sufficient supply of water and salt should be provided to workers under heat stress. The need for water supplement is of the same order as the sweat loss; 0,5-1,0 l/h may very often be required. The water should preferably contain 0,1 percent of salt to compensate for the salt lost by sweating. Salt tablets should be avoided, as overdosing may easily occur. Ideally, about one tenth of a litre of water should be drunk at a time to avoid distension of the stomach and to facilitate absorption. Adding about 100 g of glucose per litre of water might sometimes be useful to support workers on a low calorie diet, but this might increase dental caries.

As man tends voluntarily to dehydrate himself by up to 1-2 percent of his body weight, the workers should be reminded to drink deliberately. Decreasing body weight over the week and/or small volumes of highly concentrated urine indicates insufficient drinking. Dehydration of as little as 1 percent of the body weight diminishes work capacity.

Acclimatization

Time should be given for acclimatization to take place or to increase, as only about 50 percent of the possible acclimatization seems to occur spontaneously in individuals living in a hot climate. Acclimatization is an adaptive phenomenon characterized by decreasing body temperature and heart rate and increasing sweat rate at repeated work in a given situation of heat stress. Acclimatization takes place particularly during the first two days of work in a hot environment and is completed after about two to three weeks. Sodium-conserving mechanisms controlled by the adrenals also seem to be involved in acclimatization.

From the practical point of view, the time taken to acclimatize implies that full production cannot be demanded from a worker during his first week of work in a hot environment. After vacations, illness or other absences from work, a re-acclimatization period is necessary since acclimatization is lost during a couple of weeks away from work.

Heat stress disorders and first-aid treatment

Heat stress disorders show a somewhat inconsistent clinical picture, but three major conditions may be recognized:

Heat stroke

This is the most serious of the heat disorders and may prove to be fatal. The cardinal signs are the following:

(a) hot dry skin (which may be red, mottled or cyanotic - if this is possible to see for pigmentation);

(b) hyperthermia, that is, body (rectal) temperature is usually about 41° C and rising;

(c) signs of brain disturbances, such as mental confusion, delirium, loss of consciousness, convulsions or coma.

First-aid treatment should always be immediately initiated, without waiting for transport to a medical facility. First aid should begin by moving the patient to a cool place if possible, laying him down on the earth, soaking him continuously with cold water and increasing evaporation and convection by fanning. When the body temperature comes down to about 38° C, the cooling procedure should be stopped, but temperature monitoring should be continued to detect recurrent rise. Shock treatment with intravenous distribution of isotonic solutions is often necessary. (Isotonic = solutions which have the same osmotic pressure.) Among the complications of heat stroke are renal and hepatic failure, as well as myocardial impairment and brain injuries. Any one of these complications may be a threat to life, but survival is enhanced by quick and efficacious first-aid treatment.

Heat exhaustion

The symptoms are weakness, giddiness, nausea and perhaps headache. The pulse may be weak and blood pressure somewhat low, and fainting is common if the patient is standing. Body temperature is normally somewhat elevated to about 38° C; the skin is moist as sweating remains active.

The condition is not uniform in etiology but the most common causes are either water loss or a lack of salt. In the first case, there is a small volume of highly-concentrated urine and the patient is thirsty, whereas in the second case the urine is more diluted and the circulatory insufficiency more predominant. Thirst is less evident.

First-aid treatment is to move the patient to a cooler environment. Some less vigorous cooling procedures than those prescribed for heat stroke might be undertaken and dehydration is corrected by administering salted fluid by mouth. The patient should be off work and under treatment until the urinary volume is normal. This might take several days, particularly if salt depletion has occurred. Most cases recover spontaneously but medical personnel might be consulted for treatment in the more severe cases.

Heat cramps

This disorder is characterized by spasms in the skeletal muscles, for example in the legs and it occurs in people who have been sweating profusely and drunk large volumes of water without compensating for the salt losses. Treatment is to give salted water by mouth or additional salt may be added to the food in mild cases. This disorder is prevented by providing the workers with salted drinking water (0,1 percent). Salt tablets should be avoided as overdosing may occur.

It should be emphasized that salt should never be taken during work if sufficient water is not available.

Other heat disorders

Other heat disturbances have been discussed in the literature but are more or less indistinct entities and are therefore not discussed in this context. Among these are heat rash, tiny red raised vesicles on the skin. Different types of heat fatigue have also been mentioned which are phenomena depending on bad acclimatization or interactions between heat strain and socio-psychological factors.

Equipment for Medical First-Aid for Heat Stress Disorders

To provide first aid for heat disorders as outlined above, some equipment is necessary. A rectal thermometer (or an oral one - add 0,3° C to get body core temperature and avoid the thermometer being cooled by the patient's breath) is necessary to check the body temperature. Thin towels or cloths are needed to soak with water to cover the patient who is to be cooled down. (If humidity is high, it may be beneficial to use ethylalcohol in water to achieve better cooling by evaporation - other solvents are not recommended as they are readily absorbed by the skin into the body.) Also 0,1 percent salted water should be available, as well as some sort of transport for the patient. A watch showing seconds (for counting the pulse) will sometimes be useful. Thin plastic bags can be used for collecting urine for diagnostic purposes. To sum up, first aid equipment should include the following: thermometer, watch, thin towels or cloth, water, salt, (ethylalcohol), plastic bags, transport equipment.

4. High Altitudes

Physical working capacity is also reduced with increasing altitude. More specifically, the aerobic power is diminished by the reduced oxygen pressure in the air breathed in. The capacity to perform physically heavy work lasting more than two minutes starts to decrease at about 1 200 m. During acute exposure at 2 300 m, the aerobic power is impaired by about 15 percent and at 4 000 m by some 30 percent. After adaptation, the decrease in aerobic power will be cut by about half, although there is a wide range of individual variation. In forestry, those workers should be especially considered who occasionally have to perform their tasks at high altitudes while still living at low ones; as they do not have any acclimatization period, the job under these circumstances may become unduly heavy.

5. Social and Psychological Factors

Logging operations require a method of organizing production activity that is different from those employed in pre-industrial, mainly agrarian, societies. Social and psychological factors in various ways affect the production in terms of output as well as the ways in which the output can be achieved.

Social context of logging operation in developing countries

The majority of the forst workers are recruited from rural areas, mainly agricultural. They often have little or no previous experience of employment except as casual labour. Many have been self-employed.

Their basic needs are being satisfied by and through the family and other social units to which they belong. Their decisions and responses to stimuli provided by the employer - piece rates, wages and so forth - are based upon the rationality which is relevant in their context. An awareness of this will help to explain and anticipate many conflicts related to the work.

The obligations - in cash or in kind - of the worker are usually extended to cover not only his family but also the community to which he belongs. This is very important because it provides the worker with a reassurance that he in turn will be taken care of when sick, disabled or old. As long as the employer does not provide this social service, the priority of the family and the community will be dominant.

The instability of the labour force which is frequently attributed to the rural environment can thus more consistently be explained in terms of the conditions of employment. Consequently, instability of the labour force can to a great extent be reduced by the employer through improved conditions of employment.

The relation at work between fellow-workers and foremen and other categories does influence production, safety and stabilitiy of employment. Participation and partnership in decision-making, control of the immediate work situation and so on, are factors which are well known to contribute to production. The establishment of "careers" through promotion to more advanced, secure and well-paid tasks, like tractor--driving, within the labour force rather than training and employing "outsiders", is also beneficial.

6. <u>Work Motivation</u>

Work motivation may be defined as the willingness of the worker to produce and stay in the job. Work motivation is promoted if the job satisfies certain subjective needs of the individual. In behavioural science, these needs are usually grouped as follows:

(i) Needs related to physical existence, such as adequate earnings, safety of life and health;

(ii) needs related to security, such as security of employment;

(iii) needs related to self-realization and status.

In the western developed world, these needs are considered to be expressions of the individual. However, it is evident that in other cultures the prominence of the individual's need is sometimes considerably or wholly subdued in preference for the family or the tribe of which he is a member. The reaction toward or motivation in work of a worker in a developing country may therefore differ from what is usual in the developed regions.

7. <u>Cultural Traditions</u>

People in developing countries often have quite different attitudes towards work than in most developed regions in which industrialization during the last 100 to 200 years made such attitudes more uniform and predictable. The cultural traits in developing countries are very different in different localities and cannot be dealt with here. The following quotation from a paper in Ambio is chosen only as one example on the need to explore the impact of cultural traditions, including religion and taboos, in organizing work in underdeveloped parts of the world. It comes from a fuel wood plantation program in Kenya.

1. Men retain the sole right to plant, cultivate and fell trees. Women posses no such rights. They are in fact "passengers", owning nothing that might denote their permanence in the family. Their role is to bear and care for children, fetch water and cook food, and procure kuni. The last is their responsibility alone, and success is a mark of a good wife.

2. The logic behind this harsh reality is the traditional practice of land ownership. In the past, disputes over land were resolved in favor of the man who had planted (and therefore owned) the trees on the disputed land. Possession of capital assets was confined to men. How then could a woman plant a tree if she was merely a passenger? Such is the level of discrimination that taboos exist which compel the woman to stay away from tree planting. Thus, she will become barren if she plants a tree (note the link with her primary role in the household: bearing children). If she plants a tree her husband will die (i.e., she will pose a direct challenge to her husband's authority).

8. <u>Health</u>

The general level of health is usually low in developing countries and many diseases are endemic, especially in the tropics. The state of health is closely related to the diet, especially to lack of proteins and vitamins. The medical services are often also poor in rural forested areas.

Diseases are a very serious obstacle to work efficiency and measures to improve the state of health of the labour force should be included in all forest operations. Such measures will cover a wide range of items, such as improved diet, adequate housing with appropriate sanitary and hygenic facilities, medical services, clothing and provision of first aid facilities.

The following quotation from a paper by Oseni and Ward (1972) lists some of the common diseases in developing countries:

"Malaria and other fevers are common and are probably the greatest cause of loss of productivity. Sickle cell anaemia occurs when the sufferer has even a mild attack of malaria which causes the joints to ache badly and prevents him from working. It can, however, be quickly suppressed by aspirin and anti-malarial drugs taken together. Casual and other workers living mainly in villages have many days off due to sickness, particularly malaria and dysentery. Diseases, such as tuberculosis, bilharzia, filariasis, sleeping sickness, trachoma, onchocaerciasis, yellow fever, jaundice, typhoid, tetanus, smallpox and measles are also common, as are parasites such as roundworm and guinea worm. Rabies also occurs, from dogs and wild animals."

"The majority of these diseases are associated directly or indirectly with water, and consequently are most rife in villages relying upon streams for their water supply, especially as many become stagnant in the dry season. Where piped water is available, the incidence of these diseases, or of parasites such as guinea worm, is greatly reduced. Dust-borne diseases such as trachoma and meningitis are frequent in savanna areas at the height of the dry season."

"Illnesses, such as hernias, sprains and slipped discs occur because of poor carrying methods, as does hypertension caused by carrying heavy loads on the head. Other illnesses that may occur as a result of working conditions include neuroses, pneumonia, bronchitis, dermatitis and rheumatoid arthritis."

Below is a list of the commoner diseases, with their causes and comments on prevention, etc. An estimated of their frequency is also included.

Table 1:5. Common diseases in tropical countries (from Oseni and Ward, 1972).

Disease	Frequency	Cause	Comments
Anaemia	Common	Poor nutrition. Folic acid destroyed with prolonged cooking. Lack of protein	Education in nutrition
		Malaria: Destroys folic acid reserves	Eradication of malaria
		Blood loss: Hookworm and other parasites	Proper disposal of faeces - hookworm dropped in faeces enters through the feet of passers-by
Diarrhoea	Very common indeed	Bacillary dysentery: Contaminated Amoebic dysentery: water and food, flies	Proper latrines, hand-washing Boiling drinking water
		Food poisoning; Keeping cooked foods, reheating	Education in care of and cooking food
		Typhoid: Infected water supplies, direct contamination of food by hands	Proper latrines, hand washing Boiling drinking water
		For all: Washing uncooked fruit and vegetables in contaminated water	
Guinea worm	Common	Ingestion of infected cyclops (water flea). Rupture of skin and emergence of adult worm 12-18 months after infection. Worker incapacitated when worm becomes adult	If worm dies or is broken druing extraction, severe inflamation of the tissues (cellulitis), septicaemia or infective arthritis can result. Tetanus is a well-known complication of guinea worm
Hypertension	Very common in many women, fairly common in men	Two types - usual is benign, cause unknown but is thought that the carrying of head loads is a causative factor. (In young men in late teens or early 20's very serious)	Never exceed the legal limit for head loads, particularly for young people
Malaria	Endemic - all workers affected	Anopheles mosquito. Bites infected human and carries the disease on	Spraying of workers' houses, draining of stagnant water. Regular use of quinine drugs
Muscular diseases	Fairly common	Climate is an important factor	Damp clothes should not be worn; when home, change to dry clothes. Emphasis on correct methods of lifting
		Lumbago and general backache caused by climatic and lifting	

Table 1.5. (cont.)

Disease	Frequency	Cause	Comments
Ochnocerciasis	Fairly common in tropical Africa and Central and S. America	A chronic non-fatal disease caused by a nematodes carried by the simulium fly. Fibrous nodules appear on the skin. Also causes blindness	Occurs mainly from 10°-20° N in West Africa, in the savanna areas Avoid bites and control the vector
Respiratory tract infection	Very common indeed, esp. sinusites and bronchitis	Climate encourages these diseases and tropical rain forest is ideal	
Schistomoniasis (Bilharzia)	Common	A blood fluke which enters through the skin while victim is immersed in infected water Sources: water infected by contaminated faeces and urine. Certain streams and pools are known reservoirs. There are many others	Also a cause of anaemia through blood loss in stool and urine Prevention: Non-contamination of water. Limited time in and then a very brisk rub down with a rough towel of all body surfaces Health education
Tetanus	Endemic	Clostridium tetani. Reservoir is in infected animals and man. Immediate source of infection: soil, dust, animal and human faeces. Enters the body through injury, often trivial	Of importance to forestry workers who are liable to punctures and other wounds and may well be in infected areas. Prevention: Antitetanus toxoid and treatment if wounds contaminated with soil or faeces
Tropical ulcers	Common	Abrasions and cuts	Abrasions should be cleaned and covered as soon as possible (provision of good diet, washing facilities and first-aid services has abolished tropical ulcers from labour forces of well-run estates)
Tuberculosis	Very Common	Infection spread by droplets and contaminated dust and by coughing, spitting and poor housing (over-crowding - all in one room)	Health education: coughing covered - avoidance of spitting. Treatment if chest complaint suspected. Massive eradication very difficult

B. DERIVATION OF MACHINE COST

1. Depreciation

Depreciation or amortization is a means of recovering the original investment in a machine.[1] Of the various methods of calculating depreciation for costing purposes, the so-called "strait line" method based on the number of hours from service meter or recorder readings is normally used. Other methods may find their justification, e.g. in accounting procedures. The 30% on a diminishing balance method may give a truer picture as it combines obsolescence with depreciation

There is no way of knowing precisely the economic life of a machine because of factors relating to obsolescence, severity of use, quality of maintenance and so forth.

The amount to be depreciated should include import duties, sales or purchase taxes, transportation charges and all other costs incured to deliver the machine to its place of work, less the estimated trade-in or salvage value which usually taken as 10 percent of the total original cost. Depreciation per unit of time is then found by dividing this net amount by the estimated life of the machine expressed in the same time units, thus:

$$\text{depreciation} = \frac{\text{original machine cost} \cdot 0{,}9}{\text{machine life in units of time}} \qquad (1.1)$$

2. Interest

Interest for inclusion in machine operating cost is calculated by applying to the average annual investment in the machine the interest rate at which money may be borrowed to finance its purchase. It is expressed as a cost per unit of time by dividing the annual interest by the number of time units in the year, thus:

$$I = \frac{AAI \cdot i}{\text{Time units in year}} \qquad (1.2)$$

where I = interest per hour or other unit of time

AAI = average annual investment

i = rate of interest, in decimal form

---

1) Depreciate = allowance made for wear and tear
Amortize = to pay off debt through a sinking fund

While for quick and easy calculation the average annual investment is often taken as 60 percent of the delivered cost of the machine, it may be more accurately calculated with the following formula:

$$AAI = \frac{C \cdot (Y + 1) + V \cdot (Y - 1)}{2 \cdot Y} \qquad (1.3)$$

where AAI = average annual investment

C = delivered cost of machine

V = estimated salvage or trade-in value of machine

Y = estimated life of machine in years.

In some regions of the world, interest is not normally included in the operating cost record of a machine. However, it must be included when the operating costs of two or more logging machines and hence logging methods are being compared. To leave it our might lead to inaccurate machine rates and to erroneous decisions in selecting the logging method.

3. Insurance

Insurance is normally designed to cover public liability and property damage and loss of machine from fire, theft or other hazard. Its annual value is usually taken as a percentage of original machine or average annual investment and converted to a rate per machine or other time unit. The percentage rate depends on local practice but normally lies between 3 and 5 percent applied to annual average investment. The following formula may be used:

$$IN = \frac{\text{delivered cost of machine} \cdot 0{,}60 \cdot 0{,}03}{\text{units of time per year}} \qquad (1.4)$$

where IN = insurance cost per hour or other time unit

0,03 = percentage rate expressed in decimal form.

4. Taxes

This item refers to annual taxes relating to the machine proper. It includes the annual cost, if applicable, of licensing the machine, but not fuel taxes which are part of fuel cost. The annual cost of taxes is converted to a rate per hour or other time unit with the following formula:

$$\text{Taxes} = \frac{\text{annual tax amount}}{\text{average hours (or other time units) per year}} \qquad (1.5)$$

5. Operating Labour

The cost of operating labour comprises the direct wages of operator or driver and helpers used with the machine together with the indirect cost of labour fringe benefits. The latter cost must be established and expressed as a percentage of direct labour cost. Labour cost per machine hour may be found by dividing labour cost per unit period (day, week, month or year) by the number of machine hours or other appropriate time units in the period. Thus:

$$\text{Operating labour} = \frac{\text{labour cost for period} \cdot (1 + f)}{\text{PMH (or other time unit) in the period}} \quad (1.6)$$

where f = cost of labour fringe benefits expressed as a percentage of direct labour cost

PMH = productive machine hour.

6. Fuel

Fuel consumption depends on engine power and load factor. Load factor is a term used to express actual fuel consumption as a percentage of the maximum capacity of the engine to burn fuel. Thus, an engine continuously producing full rated kW is operating at a load factor of 100 percent. A load factor value of 60 percent is used often for estimating purposes.

The fuel consumption of a diesel engine using No. 2 fuel varies between 0,21 and 0,24 kg per brake kW hour, depending on engine design and speed, ambient temperature and efficiency of the engine. For calculating purposes, a mean value of 0,17 kg may be used. Fuel consumption of gasoline engines is in the order of 0,28 kg per brake kW hour.

Fuel is further discussed also in Appendix C to chapter 4.

7. Oil and Grease

The cost of oil and grease, including hydraulic oils, varies with engine kW and the capacity and complexity of the hydraulic system. For such machines as tractors, trucks, skidders and front-end loaders, which have no or relatively small hydraulic system, the cost per machine hour may be approximated with the following formula:

$$COG = \frac{GkW \cdot 0,15}{100} \quad (1.7)$$

where COG = cost of oil and grease per PMH

GkW = gross kW.

In such machines as forwarders, processors and harvesters which may have major and relatiavely complex and high pressure hydraulic systems, the cost per machine hour, expressed in cents, may exceed 50 percent and even approach 100 percent of gross kW. For example, a harvester with an engine kW rating of 200 may have an oil and grease cost of well over US$ 1 per machine hour. The cost of oil and grease for feller-buncher and knuckleboom loaders will fall between the cost for crawler tractors and that for harvesters.

Thus the formula may be expressed in this manner

$$COG = \frac{GkW \cdot x}{100} \tag{1.8}$$

where COG = cost of oils and greases per PMH

GkW = gross kW

x = factor with the following values:

(a) 0,15 for tractors, trucks, skidders, graders and frontend loaders;

(b) 0,23 for feller-bunchers and knucklebbom loaders;

(c) 0,40 for processors, harvesters and forwarders.

8. Servicing and Repairs (Except Tires for Hauling Rigs)

It is very difficult to form sound cost estimates of servicing and repairing mechanical equipment unless experience with similar machines working similar conditions is available. It has become the practice to relate lifetime servicing and repair costs to the following factors:

(a) purchase price;
(b) delivered cost less taxes;
(c) depreciation,

and to express it as a cost per productive machine hour. The third method is based on the assumption that the life span of the machine has been estimated realistically.

Although servicing and repair cost of a machine actually increases with age, it is normally averaged over machine life and expressed on a "straight line" basis in the same manner as depreciation. In practice, annual depreciataion charges decrease each year while servicing and repair costs increase, so that the two together give a fairly equal annual value over the life of the machine, thus making the "straight line" calculation reasonably realistic when both cost components are considered in this manner.

Experience shows that servicing and repair costs may vary between 75 and 150 percent of depreciation (when machine life is realistically estimated), depending on the complexity and development stage of the machine its type of work, its working milieu and the care which the operator takes. The cost consists of two items:

(a) labour, including fringe benefits
(b) parts and material.

These make up approximately 40 percent and 60 percent respectively of the total cost.

In developing countries, the cost of parts and material often differs substantially from that in industrialized countries, but labour cost is much lower. However, lower mechanic training and efficiency in developing countries would offset the lower wage structure to some extent. As a general rule lifetime servicing and repair cost of a reasonably well developed logging machine is considered to equal the delivered cost of the original machine.

$$\text{Servicing and repairs} = \frac{\text{delivered cost of machine}}{\text{normal life in hours}} \qquad (1.9)$$

9. <u>Tires (for Hauling Units Only)</u>

Tires of hauling rigs are sometimes costed on a mileage basis. There is nothing wrong with this method, particularly where the trucks are operating continuously under the same loads and/or over the same type of roads. However, on most logging operations, including those in plantations, several types of roads are encountered from rough low-speed roads where tires are often subject to sidewall damage, to smooth, paved high-speed roads where hazards are few and heat is probably the greatest tire enemy. It seems obvious therefore that tire costing on the basis of travelling hours will be as reliable as or even more accurate than costing on a mileage basis. It is also more easily used in costing procedures.

It is also obvious that experience with similar equipment and tires on similar operations will provide the best information on tire cost. When this is not available, other means must be found. Much will depend on construction of the tire, tire load and inflation pressure, road surface, travel speed, ambient temperature and numerous other factors.
The original tires should be depreciated or written off with the truck or trailer in the case of log hauling rigs. Thus, tire costs will cover repairs to the original tires and the cost of and repairs to replacement tires during the life of the unit. This may be found by evaluating the following expression:

$$\frac{B}{Y \cdot Z} + \frac{(T + B)\ (Y \cdot Z - A)}{A \cdot Y \cdot Z} \qquad (1.10)$$

where B = life repair cost of a set of tires

T = replacement cost of a set of tires

Y = life of truck or trailer in years

Z = travelling hours per year

A = life of a set of tires expressed in travelling hours

$Y \cdot Z$ = lifetime travelling hours of truck or trailer

$\frac{B}{Y \cdot Z}$ = lifetime repair cost of original tires

$\frac{T + B}{Y \cdot Z}$ = cost of replacement set of tires and repairs to those tires spread over machine life

$\frac{Y \cdot Z - A}{A}$ = number of sets of replacement tires required during machine life.

On the rough basis that (a) a hauling rig should spend 10 000 h of its life travelling: (b) tire life should be about 2 000 travelling hours; and (c) recaping and repairs per tire amount to 50 percent of the original cost, four replacement sets of tires will be needed during the lifetime of the hauling rig at a cost of

$$CST \cdot 1\ 5 \cdot 4$$

and tire cost per travelling hour will be found with the folloving formula:

$$TC = \frac{CST \cdot 1{,}5 \cdot 4}{10\ 000} = 0{,}0006 \cdot CST \qquad (1.11)$$

where TC = tire cost per travelling hour

CST = cost of set of replacement tires.

The labour cost of changing tires and carrying out minor tire repairs may be, for convenience sake, included with servicing and repair labour.

C. NOTE ON ACCOUNTING PRICES OR SHADOW WAGE RATES FOR LABOUR AND LAND IN FORESTRY DEVELOPMENTS

It is now generally recognized that market prices of inputs (e.g. labour and land) in the making of cost-benefit evaluations of forestry developments do not always reflect the real value or the true costs for the economy of these inputs. Consequently, evaluations of net benefits for the economy may be untrue or distorted.

Economists generally agree on the principle behind the use of "accounting prices" or "shadow prices" instead of market prices in order to measure real values for the economy in cost-benefit estimates and to rank development opportunities correctly. However, various ways are suggested for the derivation of such shadow prices. Two of the most commonly used are described in detail in the references (1) and (2). Only a few remarks will be made here with regard to the level of shadow prices in relation to market prices.

1. <u>Labour</u>

In principle, the shadow rate for labour is equal to its opportunity cost[1] derived from the marginal productivity of labour in the sector from which it is drawn. The level will consequently depend on the real value of the marginal product in that sector. If this value is low, as for example in subsistence farming, the shadow wage rate may often be in the bracket of 40% of the market wage rate. If in a theoretical case, no other employment opportunity than forestry exists, the shadow wage rate would be zero (plus possibly the value of the increased consumption of the individual caused by his work in forestry). In an economy with full employment the shadow wage rate will in theory be equal to the market wage rate, whereas under shortage of labour it will exceed the latter. It has been suggested that in a forestry region with not too severe unemployment or under-employment, the appropriate shadow wage rate would be around 75% of the market wage rate.

2. <u>Land</u>

If there is no alternative use of land than forestry, the shadow price of the land is zero. Such a situation may seldom exist as some benefits - many of them perhaps not measurable - may often accrue from any piece of land. If an alternative land use exists, the shadow price for forestry is derived on the basis of the social values of the benefits of this alternative land use.

---

1) Opportunity costs = costs that are foregone or attributable to doing one thing rather than another - implied costs that accrue to factors which might be made in alternative ways.

3. Conclusions

There are some serious difficulties both in deriving and applying shadow prices, nevertheless, shadow pricing is now in rather commmon use in many countries in cost-benefit evaluation of development projects. It is important that also forestry projects are evaluated on the same basis so that they are ranked properly and their social feasibility is properly evaluated relative to other sectors. The determination of a shadow wage rate for labour is particularly important in the social evaluation of forestry projects since they are often labour-intensive. Even small variations in the shadow wage rate may have a considerable effect on the final results of an evaluation of plantations and other silvicultural operations because of the effect of discounting over the long time periods required to grow a forest crop.

References

Baumol (1969) has written a classical article dealing with accounting prices. For a general treaty see e.g. Irvin (1978) or Dasgupta (1978) or papers issued by international aid agencies such as The World Bank.

(1) OECD: Manual of Industrial Project Analysis, Vol. II Social Cost/ Benefit Analysis, by J.M.D. Little and T.A. Mirrless (Paris 1969).

(2) UNIDO Guidelines: Alternative Project Selection Procedures for Developing Countries, by D. Dal (IBRD memo April 1973).

(3) Leslie, A.J.: Economic Problems in Tropical Forestry. FAO, Rome 1971 (FO:Misc/71/24).

(4) Watt, G.R.: The Planning and Evaluation of Forestry Projects. Institute Paper No. 45, Commonwealth Forestry Institute, University of Oxford, 1973.

(5) Layard, R. (Ed.): Cost-Benefit Analysis. Penguin Modern Economic Readings, 1972.

(6) Pearce, D.W.: Cost-Benefit Analysis. MacMillan Studies in Economics, London and Basingstoke, 1971.

(7) Baumol, W.J.: On the Discount Rate for Public Projects. In The Analyses and Evaluation of Public Expenditures: the PPB System. Vol. I, U.S. Congress, Washington D.C. 1969, pp 489-504.

(8) Irvin, G.: Modern Cost-Benefit Methods: An Introduction to Financial, Economic and Social Appriasal of Developments Projects. MacMillan Press Ltd, London 1978.

(9) Dasgupta, A.K. and Pearce, D.W.: Cost-Benefit Analysis: Theory and Practise. The MacMillan Press Ltd, London 1978.

APPENDIX TO THE CHAPTER 2. PROBLEM ANALYSIS

A. THE GEOMETRY OF FOREST TRANSPORT

B. CHECKLIST FOR LOGGING, TRANSPORT AND ROADING CONDITIONS

A. THE GEOMETRY OF FOREST TRANSPORT

Forest transport seldom follows the straight line. But the straight line distance, often in the ground plane, is a suitable concept in transport studies. Allowance is then made for the lengthening by curves and difference in altitudes by a winding coefficient, e.g. (real) transport distance = straight line distance · winding coefficient

or

$$S_t = S_s \cdot w \qquad (2.1)$$

Some examples are worth elaborating. If A and B, the terminals of the transport, are located at the both ends of the hypotenuse of a right-angled triangle and the transport follows the legs the winding coefficient takes the value 1,4 (or exactly √2). It does not change if the transport follows the small sides of a series of smaller triangles as drawn in the figure 2.1. A winding coefficient of 1,4 means a 40% longer route than the straight line.

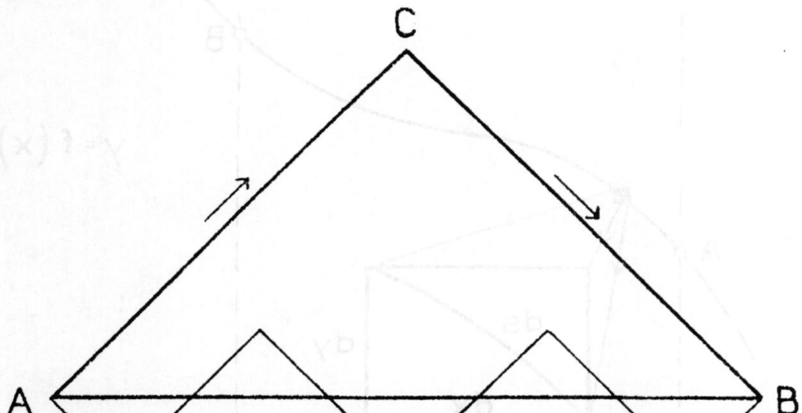

Figure 2.1. The winding coefficient for a path following the legs of a right-angled triangle is close to 1,4 (=√2)

If A and B are located at the ends of the diameter of a circle the winding coefficient takes the value of

$$\frac{\pi}{2} \approx 1,6$$

Also this value is not affected if the transport follows the periphery of a series of half-circles as in the figure 2:2.

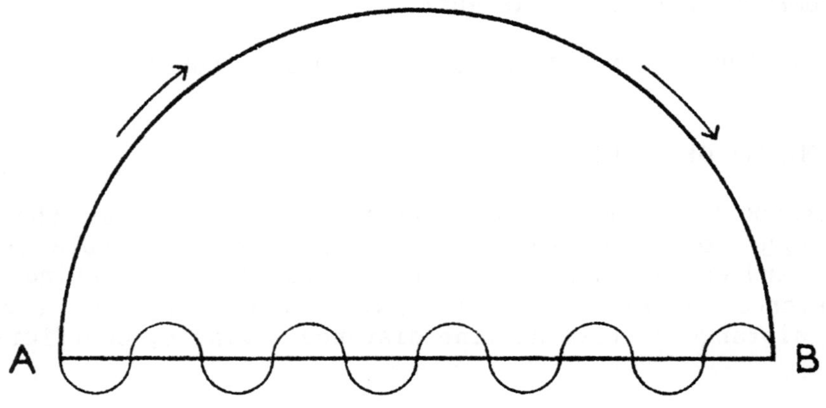

Figure 2.2. The winding coefficient of a path following the periphery of a circle is close to 1,6 (=π/2)

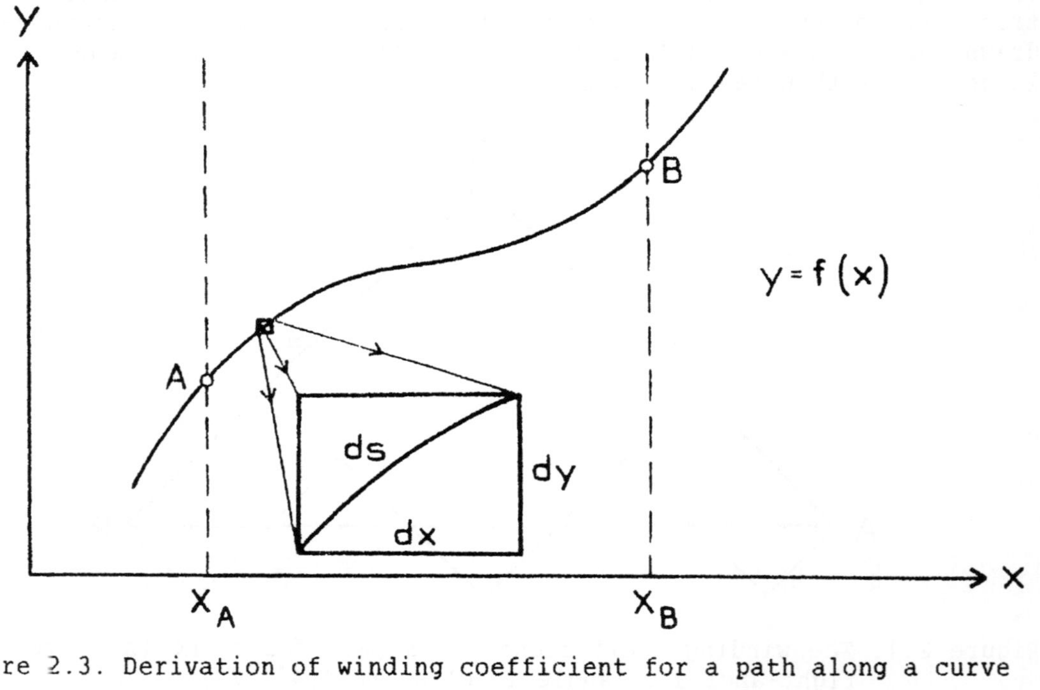

Figure 2.3. Derivation of winding coefficient for a path along a curve

If section AB of a route (see figure 2.3) follows the function y = f(x), the route distance S (= the length of the curve AB) can be calculated as follows:

Consider a small section of the curve. Then, according to Pythagoras' theorem

$$(ds)^2 = (dx)^2 + (dy)^2 \text{ and} \tag{2.2}$$

$$\frac{ds}{dx} = 1 + \frac{dy}{dx}^2 \tag{2.3}$$

By limit analysis is derived

$$S' = 1 + (y')^2 \tag{2.4}$$

and the curve length becomes

$$S = \int_{x_A}^{x_B} \sqrt{1 + (y')^2}\, dx \tag{2.5}$$

A special case in applying this formula is when A and B are located at the ends of a sinus curve of "standard form". Such a curve crosses the line AB with an angle of 45° and the height of the curve is 1 when the length AB is 2π. For this particular case the route along the sinus curve, S, becomes

$$S = \int_0^{\pi/2} \sqrt{1 + \cos^2 x}\ .\ dx = \sqrt{2} \cdot 1{,}3506 \tag{2.6}$$

and the winding coefficient

$$\frac{S}{AB} = \frac{4 \cdot \sqrt{2} \cdot 1{,}3506}{2\pi} = 1{,}216 \tag{2.7}$$

The winding coefficient, which always exceeds 1,0 varies for different modes of transport. It usually takes a higher value for lower road standards and decreases for more developed modes of transport.

Many times there is a need to calculate the average transport distance from an area to a road. Two situations will be explored here. In the first one, it is assumed that each piece of cargo is carried to the nearest point of the road. The transport route extends perpendicular on the road and the goods are piled up continously along the road. (This is a rather unrealistic situation but it is nevertheless of interest from a more theoretical point of view.) In the second situation the goods are carried to certain centres or points along the road, e.g. landings.

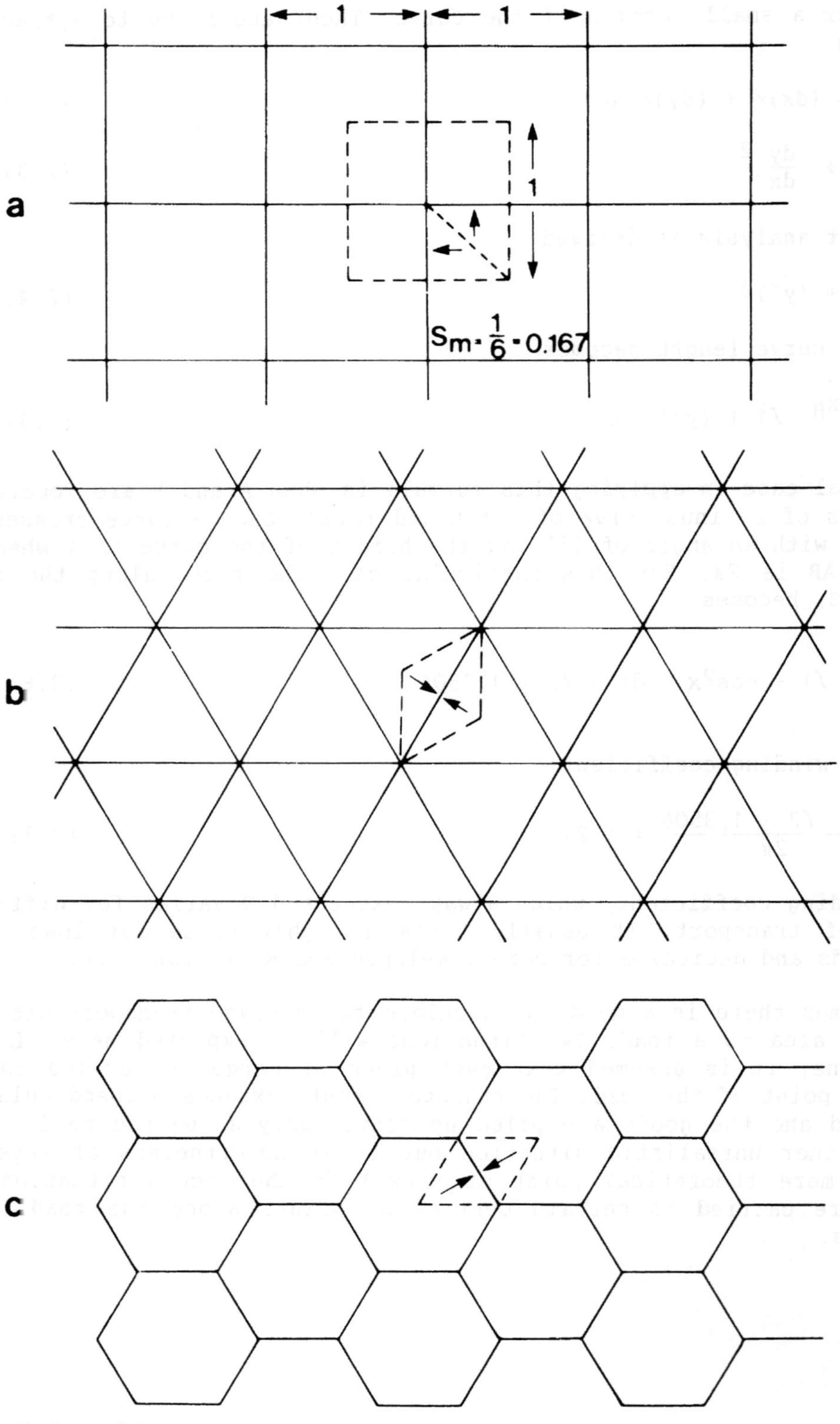

Figure 2.4. Uniform patterns of roads
(a) infinite parallel = rectangles with infinite side
(b) square
(c) triangle
(d) hexagon

In the chapter 2.5 four uniform patterns have been studied - (i) infinite parallel lines, (ii) triangular, (iii) squares and (iv) hexagon (see the figure 2.4, a-d, p. 180). If we now let the lines symbolize roads it is evident that the average transport distance fo road pattern (i) - infinite parallel roads - is 1/4 of the spacing between the roads. For the other patterns (ii) - (iv) the calculation is based on the fact that the average transport distance from a triangle to one of its sides is 1/3 of the height (against this side) of the triangle. As the area now can be subdivided in uniform triangles with the road as one side, it is only a matter of assessing the height of this triangle in relation to its base side (which is the road).

Let us as an example take the square pattern. If the side of the square is 1, the height of triangle is 1/2, the average transport distance 1/3 · 1/2 = 1/6 and the road density 4 unit length per unit area (see figure 2.5).

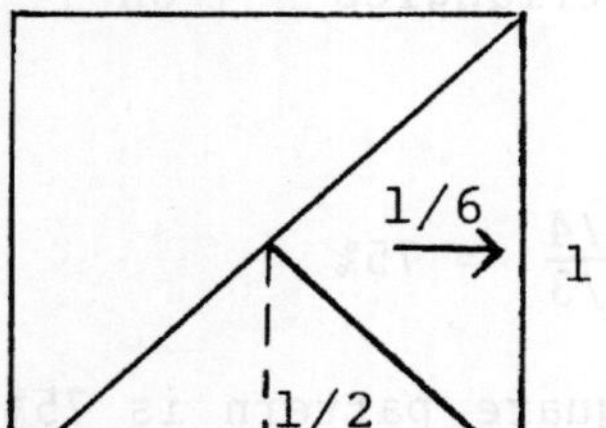

In comparing the efficiency of a road net in square pattern with infinite parallel roads, we find that we must blow up the scale of the squares to the double size to arrive at the same road density.

Figure 2.5. For a square with the side 1 the average transport distance to its sides is 1/6

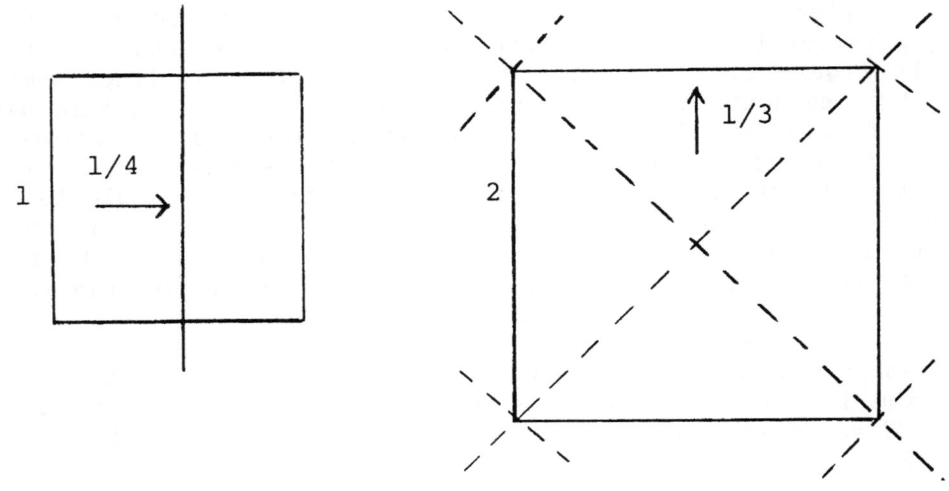

Road density 1 on 1 — 2 on 2 triangles = 1 on 1

Are transp. distance 1/4 — 1/3

Efficiency of road net 100% — $100 \cdot \frac{1/4}{1/3} = 75\%$

Figure 2.6. The efficiency of roads in square pattern is 75% of infinite parallel roads

Similar calculations for the patterns of triangles and hexagons result in a road efficiency of 75% also for both these cases.

Let us now reverse the problem by letting the lines of the square, triangle and hexagonal represent boarders of area figures. We further assume a transport flow from the figure to its centre. This is a situation when the goods from the area shall be collected to a central point, say sawlogs to a sawmill. This calculation is best based on the formula for the average distance of a triangle with two equal sides (see the figure 2.7) to its top (where the equal sides meet).

$$s = \frac{1}{3}\left[\sqrt{(a^2 + h^2)} + \frac{h^2}{a} \cdot {}^{e}\log \frac{a + \sqrt{(a^2 + h^2)}}{h}\right] \qquad (2.8)$$

We shall now elaborate two special cases of this general formula:

(i) for a right angled triangle with the base 1 (2a = 1) the formula takes the value

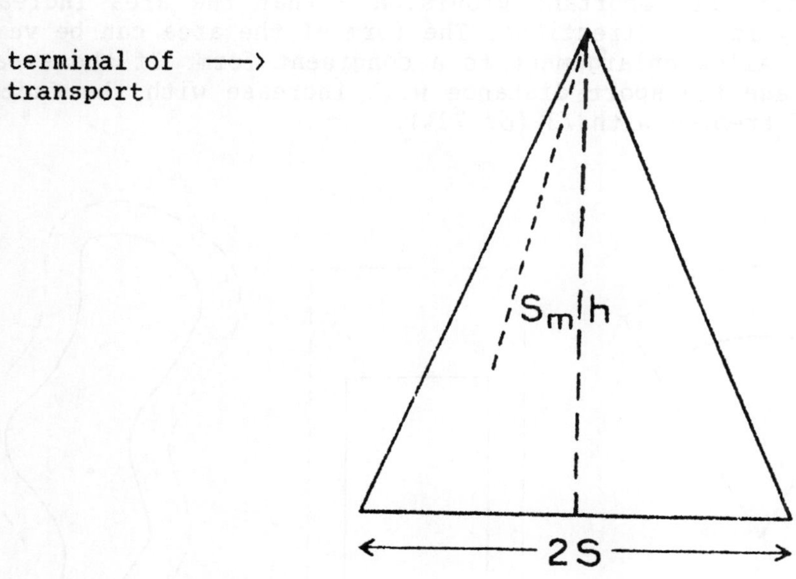

Figure 2.7. Deriving transport distance for a triangle with two equal sides

$$s = \frac{1}{3}\left[\sqrt{2} + {}^{e}\log\,(1 + \sqrt{2})\right] = 0{,}3826 \qquad (2.9)$$

This represents the square pattern for the case when the area of the square is 1.

(II) for a equilateral triangle (with three equal sides), representing the hexagon (which can be divided into six such triangles), the formula takes the value 0,3772, provided the area of the hexagon is 1.

$$s = \frac{1}{3\sqrt{3}} \cdot \left(1 + \frac{3}{4} \cdot {}^{e}\log\ 3\right) = 0{,}3772 \qquad (2.10)$$

From this we can conclude that to obtain the shortest possible average transport distance from an area to a point in its centre, this area should have the form of a hexagon. We have now assumed that this area is a subunit for a larger region and that all these equal sub-units shall give full coverage.

The average distance for a circle or a sector of a circle is 1/3 of the radius. But circles put together do not give full coverage for an area.

Another conclusion drawn from studies of this kind is that if we want to increase the area for a point of collection with the factor X, the average transport ditance will increase with the factor $\sqrt{X}$, however, only under one important provision - that the area increases proportionately in all directions. The form of the area can be very irregular but must allow enlargement to a congruent form. If the area is doubled the average transport distance will increase with the factor $\sqrt{2}$ (or ≈ 40%), if trebled with $\sqrt{3}$ (or 73%).

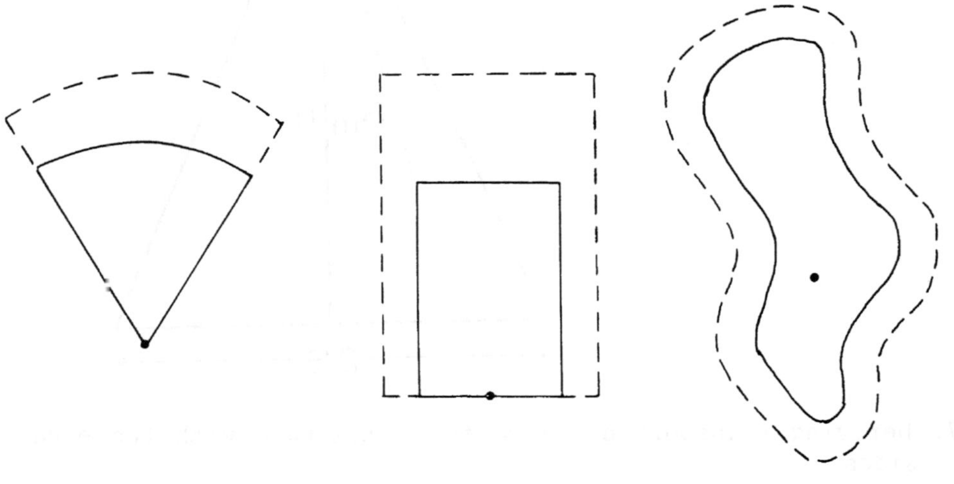

Figure 2.8. Logging areas of congruent form from one point (e.g. landing)

Examples of congruent figures for which the average transport distance increase with $\sqrt{\frac{\text{larger area}}{\text{smaller area}}}$ are shown in the figure 2.8.

Area figures or road net patterns are not in real life as nice as those dealt with up to now. We therefore need additional tools for our assessment. Forest areas can often be simplified to rectangles without violating the real shape too much. And each such rectangle can in turn be subdivided into smaller rectangles of many shapes with regard to size and ratio short/long side. In order to elaborate such situations we use the rather comlex formula for the average transport distance for a rectangle to one of its corners

$$s = \frac{1}{3}\sqrt{(a^2 + b^2)} + \frac{a^2}{6 \cdot b} \cdot \text{logtg}\left(\frac{\pi}{4} + \frac{\varphi}{2}\right) - \frac{b^2}{6 \cdot a} \cdot \text{logtg}\,\frac{\varphi}{2} \qquad (2.11)$$

where $\varphi = \text{arctg}\,\frac{b}{a}$ and a and b are the sides of the rectangle. This

rather cumbersome formula is tabulated in the table 2.5 on p 64 for a number of locations of the point of collection. Almost any configuration of rectangular shape can be solved approximately with the aid of this table but a subdivision of the rectangle in a number of smaller ones may often be necessary. The distances of the smaller rectangles are then aggregated by weighting.

In cases when the area figures cannot be simplified to rectangles we need still another method in order to master fully irregular figures. This method is based on the fact that the average distance of a circle to its origin is 2/3 of its radius. This also holds for a sector of a circle. Each irregular figure can now be subdivided into a number of sectors of a circle with its centre at the point of collection. The larger the number of such sectors, the more precise will the assessment be. If all the sectors $y_1 \ldots\ldots y_i$ have the same area the average distance becomes

$$s = \frac{2}{3} \cdot \frac{\Sigma(r_i \cdot y_i)}{\Sigma y_i} \qquad (2.12)$$

which can be put in the more useful form

$$s = \frac{2}{3} \cdot \frac{\Sigma r_i}{N_i} \qquad (2.13)$$

N = number of sectors

However, to construct a large number of sectors of equal area but with varying radii is a very time consuming procedure. The use of equiangular sectors is recommended. The formula then takes the form

$$s = \frac{2}{3} \cdot \frac{\Sigma r_i^3}{\Sigma r_i^2} \qquad (2.14)$$

In practise the procedure will be the following. Construct a template of some transparent material of the type shown in the figure 2.9, a. Put this templet on the point of collection and read the length of each radius from origin to the perimeter of the area. This distance is "r_i" in the formula above. However, if the area is so irregular that three intersections occur between a radius and the perimeter, all three

r_i's shall enter the formula but the middle one, r_2 in the figure 2.9, c, shall enter the formula with a negative value both in the numerator and in the denominator. Thus, also $r_2{}^2$ shall be negative in the summing up of the $r_i{}^2$ in the denominator.

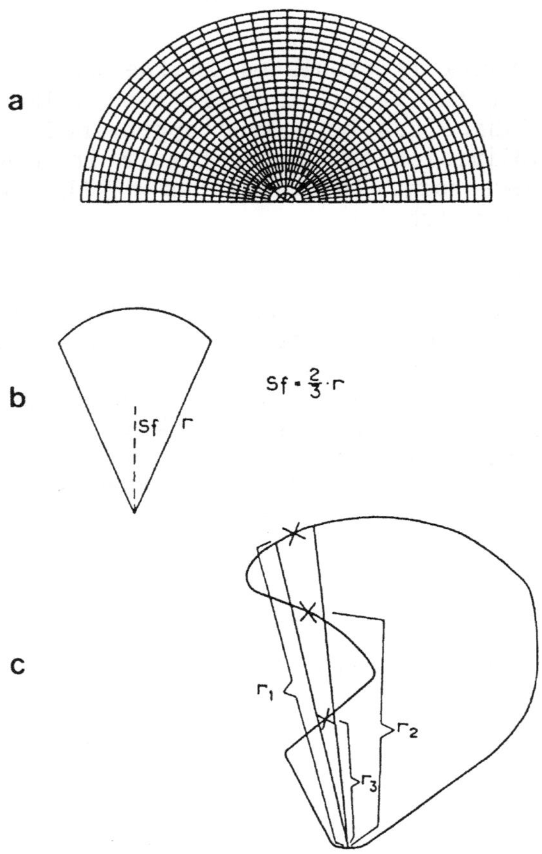

Figure 2.9 (a) Template of transparent material
(b) Average transport distance from a sector to its origin is two thirds of the radius
(c) Irregular area where roadline has three intersections (X) with the boundary

Another common configuration of an supply area is shown in the figure 2.10, circle 1 and we want to know the numerical value of R as compared to r, circle 2, for a full circle with equal area, y. Then

$$\frac{y}{a^2} = \left(\frac{R}{a}\right)^2 \cdot \left[\frac{\pi}{2} + \arcsin\frac{a}{R} + \frac{a}{R}\sqrt{\left(1 - \frac{a^2}{R^2}\right)}\right] \qquad (2.15)$$

transformed to

$$\frac{r}{a} = \frac{R}{a} \cdot \sqrt{[\frac{1}{2} + \frac{1}{\pi} \cdot \arcsin \frac{a}{R} + \frac{a}{\pi R} \sqrt{(1 - \frac{a^2}{R^2})}]} \qquad (2.16)$$

The procedure will be, assuming that Y is assessed, to

1. determine $r = \sqrt{y/\pi}$
2. compute r/a
3. read R/a in the graph, figure 2.10, which is based on formula (2.16).

The average transport distance to the centre of circle 1 can then be estimated by weighting of the distances for the two areas, the sector of the circle and the triangle:

1. For the sector it equals 2/3 R
2. For the triangle - assessed with the formula (2.9).

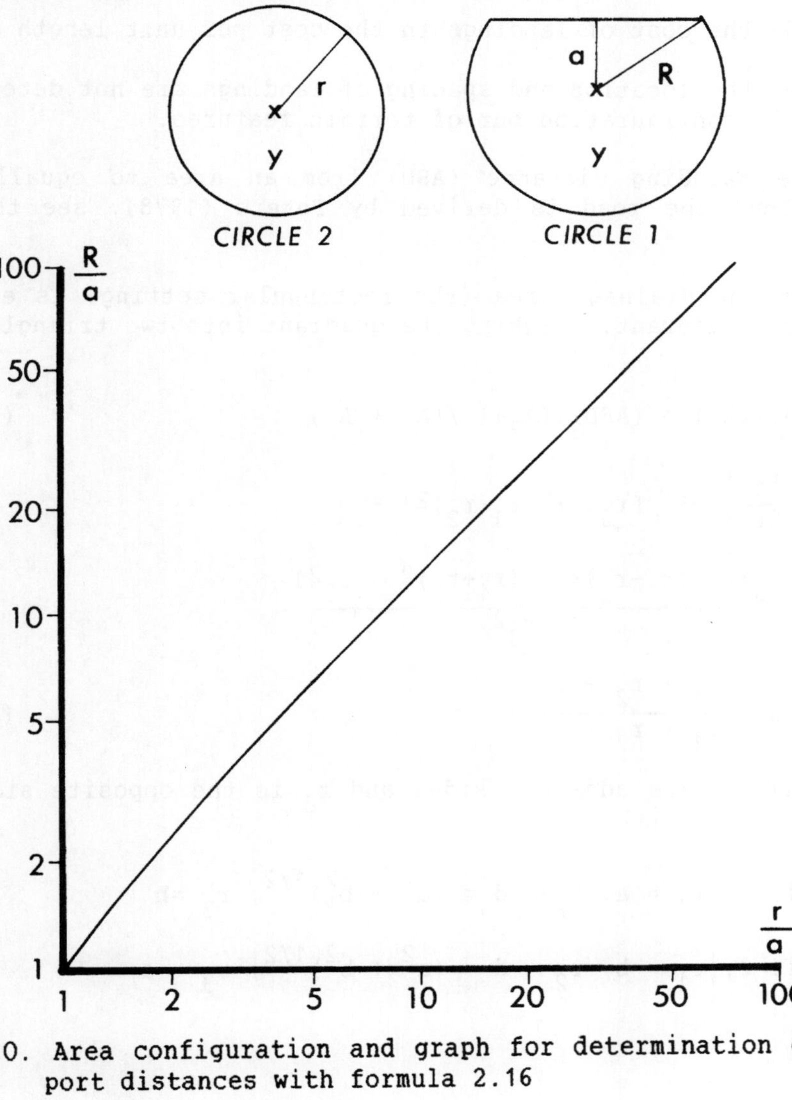

Figure 2.10. Area configuration and graph for determination of transport distances with formula 2.16

The optimum spacing is obtained when the variable extraction cost equals the road cost. Peters (1978) also analysed the inference of the landing cost on the optimum spacing and concluded that road and landing spacings are optimized when the variable extraction cost equals the road cost plus twice the landing cost.

To achieve simultaneously optimum road spacing and landing spacing according to Peters' method involves rather complex mathematics, for which Peters works out special tables as an aid. It seems that his solution will conform in principle with the more simple but approximate method of using formula (2.25), p. 83 provided that the following is observed.

(i) Include the T factor in the formula whereby the increase in extraction distance and cost by delivery to landings instead along the road is accounted for

(ii) Include the cost of landings in the cost per unit length of roads.

In practice, the location and spacing of landings are not determined by the geometric configuration but of terrain features.

The average skidding distance (ASD) from an area to equally spaced landings along the road is derived by Peters (1978), see the figure 2.11.

The ASD for the drainage area (the rectangular setting) is equivalent to that for a quadrant. Breaking the quadrant into two triangles we can write

$$ASD = [(ASD_1)(A_1) + (ASD_2)(A_2)] / (A_1 + A_2) \qquad (2.17)$$

$$\text{The ASD} = \frac{r_1 + r_2}{6r_3^2} [r_3^2 + (r_1 - r_2)^2] + \frac{[r_3^2 - (r_1-r_2)^2]\,[(r_1+r_2)^2 - r_3^2]}{12\, r_3^2} \cdot \ln \left[\frac{r_1 + r_2 + r_3}{r_1 + r_2 + r_3} \right] \qquad (2.18)$$

where r_1 and r_2 are adjacent sides and r_3 is the opposite side of any triangle.

For triangle 1; $r_1 = a$, $r_2 = d = (a^2 + b^2)^{1/2}$, $r_3 = b$

For triangle 2; $r_1 = b$, $r_2 = d = (a^2 + b^2)^{1/2}$, $r_3 = a$

Assuming a = S/2 and b = L/2, the ASD for the rectangular setting can then be transformed to:

$$ASD = 1/6 \{(S^2 + L^2)^{1/2} + (L^2/2S) \ln [(S + (S^2 + L^2)^{1/2})/L] +$$

$$+ (S^2/2L) \ln [(L + (S^2 + L^2)^{1/2})/S]\} \qquad (2.19)$$

Since this equation is rather cumbersome for practical use, the influence of landings can in practical application be included in the T-factor (see chapter 2).

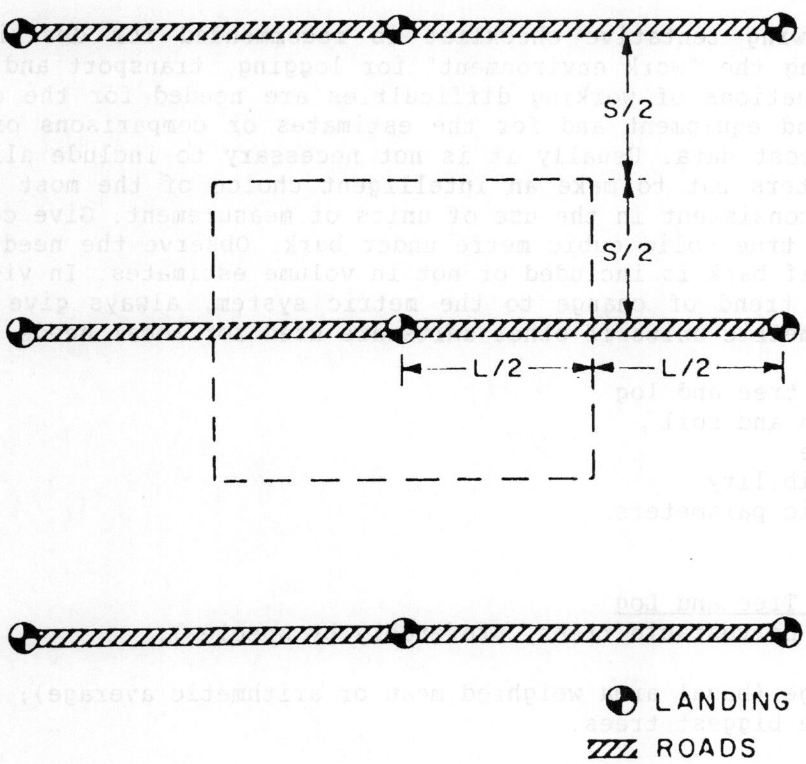

Figure 2.11. Geometry of rectangular area, two-way yarding. L = landing spacing and S = road spacing

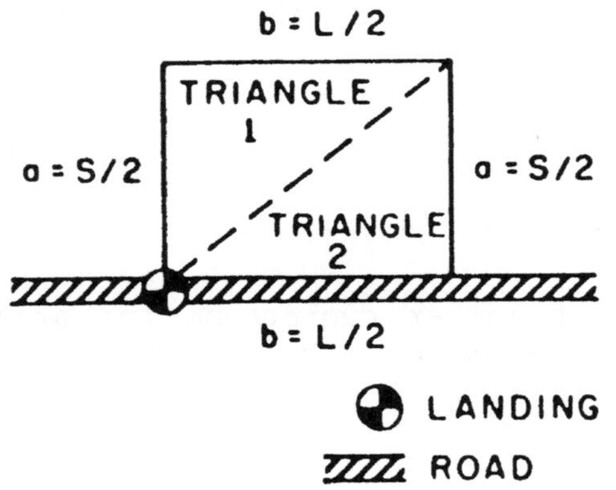

Figure 2.12. Upper right-hand quadrant of rectangular harvest setting. L = landing spacing and S = road spacing

B. CHECELIST FOR LOGGING, TRANSPORT AND ROADING CONDITIONS

The following tentative checklist is recommended for describing and quantifying the "work environment" for logging, transport and roading. Such evaluations of working difficulties are needed for the choice of methods and equipment and for the estimates or comparisons of production and cost data. Usually it is not necessary to include all factors or parameters but to make an intelligent choice of the most important ones. Be consistent in the use of units of measurement. Give conversion factor to true solid cubic metre under bark. Observe the need of being specific if bark is included or not in volume estimates. In view of the universal trend of change to the metric system, always give main results in metric units if other units are used for local needs.

1. Stand, tree and log
2. Terrain and soil
3. Climate
4. Accessibility
5. Economic parameters
6. Labour

1. <u>Stand, Tree and Log</u>

<u>DBH</u> (of harvestable trees, preferably by species):

(a) average (basal area weighted mean or arithmetic average);
(b) of the biggest trees.

<u>Note</u>: State smallest tree included in the estimate.

<u>Stump diameter</u>:

At felling cut (can be given as ratio of DBH). For tropical hardwoods indicate occurrence and mean height of buttresses.

Height:

(a) total, of dominant trees or related to some DBH parameter;
(b) for hardwoods also: clear bole height or height to given diameter.

Branchiness:

Length of green crown, preferably related to some DBH parameter. Occurrence of dead branches may also be quantified.

Bark:

Bark thickness, characteristics and volume.

Other tree parameters:

Straightness of bole, frequency of defects, rot, etc.

Volume per unit area:

(a) merchantable or presently utilized volume;
(b) total volume or other indication on the density of the stand (number of trees per hectare, etc.).

Undergrowth:

Description.

Log size:

(a) average volume or weight of logs;
(b) diameter and length specifications (maximum, minimum, standard length, etc.).

Note: Specify where boles or logs are crosscut.

Wood density:

Oven-dry density and green density.

Other log parameters:

Bark, straightness, maximum decayed wood allowed.

2. Terrain and soil
Topography, terrain pattern:

Describe geomorphological features such as relief amplitudes, regular or irregular drainage pattern, stream frequencies, rock outcrop, etc.

Terrain unit (Area of a single physiographic feature):

Describe dimension of terrain unit by relief amplitude, length and width.

Rivers and creeks:

Width and depth of streams to be passed in skidding and roading.

Terrain component:

(a) Slope: Measure slope in percent rather than in degrees. Disregard slopes shorter than 50 m (included in gullies and ground roughness). State how slope measurements and classifications are made (field measurements, on topomaps with scale and contour intervals, sampling, etc.).

(i) Indicate main slopes longer than 50 m. Disregard flattening sections of the slope nearest to ridge and valley bottom. Indicate usual length of slope and classify hillsides in convex, concave, s-formed or combination of these.

(ii) Break down the area, if possible, into the following slope classes:

Slope class	Percent of area
<10%	
10-33%	
33-50%	
>50%	

(iii) Break down the area steeper than 50 percent, if possible, into the following slope length classes:

Length of slope > 50%	Percent of area
50-300 m	
300-700 m	
>700 m	

(b) Microtopography. Ground roughness:
Frequency of gullies and slopes shorter than 50 m. Frequency of obstacles with a relief amplitude exceeding 0,5 m.

(c) Areas Flooded:
State cause (precipitation, tidal water, etc,), duration and frequency.

Soil:

Limit the classification to the following soil properties:

(a) Bearing Capacity for off-the-road vehicles: in moist state
Poor - silt, loam and organic soil
Good - other soils (friction soils).

(b) Suitability for roading:
   (i) Suitable or unsuitable for sub-base of truck road
   (ii) Suitable or unsuitable for surfacing of truck road
   (iii) Depth of soil, rock outcrops.

If the soil is not suitable for roading, indicate distance to gravel pits, river bed gravel, etc.

(c) Susceptibility to erosion:
Describe risk of erosion on logging areas and for road construction.

3. Climate
Use significant parameters in available records, such as:

Rainfall:

Amount and distribution of annual rainfall. Maximum recorded rainfall per hour and day.

Water discharge and its variation

Temperature:

Monthly averages, maximum and minimum.

Insolation:

Length of sunny periods during rainy seasons.

4. Accessibility
Access from major markets:

(a) distance to wood-using centres (mills, cities, densely populated areas, export ports);

(b) existing transport routes (rail, road, waterways), their standard and capacity;

(c) relevant traffic legislation.

Roadnet inside the perimeter of the forest area(s):

(a) density of roads, preferably in metre per hectare;

(b) breakdown of forest roadnet on standards;

Alternative 1 (rough classification)

Type of Road	Average roundtrip speed of travel for timber trucks, km/h			% of Roadnet
	Flat/undulating	Hilly	Steep	
Artery Roads	45	35	25	%
Primary Roads	25	20	15	%
Secondary Roads	20	15	15	%

Alternative 2 (classification related to specified timber trucks)

Type of road	Topo-graphy	Speed (km/h)			Type of timber truck			
		Loaded	Empty	Aver.	GVW tonne	Payload tonne	kW	No. of axles
...	...	...	...	...	...	...	...	...
...	...	...	...	...	...	...	...	...
...	...	...	...	...	...	...	...	...
...	...	...	...	...	...	...	...	...

(c) breakdown of forest road net on availability: indicate what portion of the roadnet is not suited for use during periods of rain. State length of such periods.

Extraction (skidding) distances:
If road density is assessed, the average extraction (skidding) distance can approximately be estimated with the formula:

$$\bar{S} = \frac{a}{D}$$

$\bar{S}$ = average extraction (skidding distance, in kilometres)

D = road density, in metre per hectare

a = "Spatial" constant, equals

4 to 5 for flat-undulating terrain
5 to 7 for hilly terrain
7 to 9 for steep terrain
9 or more for very steep terrain

Note: the estimate is asuming that the timber crop is homogeneously distributed over the area.

5. Economic parameters

(a) Exchange rate of money (local currency to US$), inflation rate - past and expected.
(b) Cost of equipment - purchase prices (indicate date) for standard logging equipment, cost of fuel, facilities for servicing and so on.

(c) Give main cost data also in a well known currency such as US$.

6. Labour

(a) Labour availability and skill, transport and camp facilities.

(b) Number of workdays per year, effective working hours per week and day.

(c) Payment systems, wages and fringe benefits for different categories of workers, for example:

- ordinary labour;
- skilled/specialized labour (subdivided in suitable groups, such as operators of power saws, light and heavy tractors/trucks, etc.);
- foremen/work supervisors/technicians;
- professionals.

APPENDIX TO THE CHAPTER 3. APPLICATIONS AND QUANTIFICATIONS

A. EXAMPLE OF THE RATIO $RC_n:Q_i$ AT DIFFERENT AGE DISTRIBUTIONS AND PLANNING PERIODS (REFERENCE TO TEXT ON PAGE 110-114)

B. SOME CURVE FORMS AND FUNCTIONS

A. EXAMPLE OF THE RATIO $RC_n:Q_i$ AT DIFFERENT AGE DISTRIBUTIONS AND PLANNING PERIODS (REFERENCE TO TEXT ON PAGE 110-114)

The following conditions applies for this example:

(i) Rate of interest, 5%
(ii) Annual road maintenance cost 3% of road construction cost
(iii) Stand management: 85 year rotation with 3-5 thinnings yielding 45% and final cut 55% of total yield
(iv) Thinnings accumulated (neglected) one treatment interval, awaiting road extension to the area

The ratio $\frac{RC_n}{Q_i}$, as computed using the formula (3.13) on p. 109 and (3.14) on p. 110 is shown in the graph, figure 3.1.

Alternative A on the horisontal axis is an even aged distribution, to the right 100% mature forests for clear cut (G), to the left (F) only newly established forests. D,C,B,E are intermediate age distributions. The numeraire on the vertical axis is directly proportional to the road spacing to be understood as follows: Lowest numeraire (= closes spacing) is obtained at agedistribution G, the whole area with mature stand to be clearcut. Moving to the left from G through E, B to the even distribution A the numerarie increases only slightly. For F at the far left the numeraire is 3-4 times of its value for G. The length of the planning period is only of practical importance for the age distributions F,D and C for which young (less than half the rotation age) stands dominate in the area.

Looking upon road planning for a forest as a dynamic problem in time perspective the following conclusions can be made:

A complete road net shall be built prior to clearcuts of at least one fifth of the area. If the actual age distribution makes this possible a planning, period of around 1 year (= for the immediate logging operation) ca be used. If an area has only newly established stands, then the optimum road spacing shall be 3 to 4 times wider. This road net shall then be extended as these stands reach maturity and are ready for clear cutting. (See further Sundberg, 1963.)

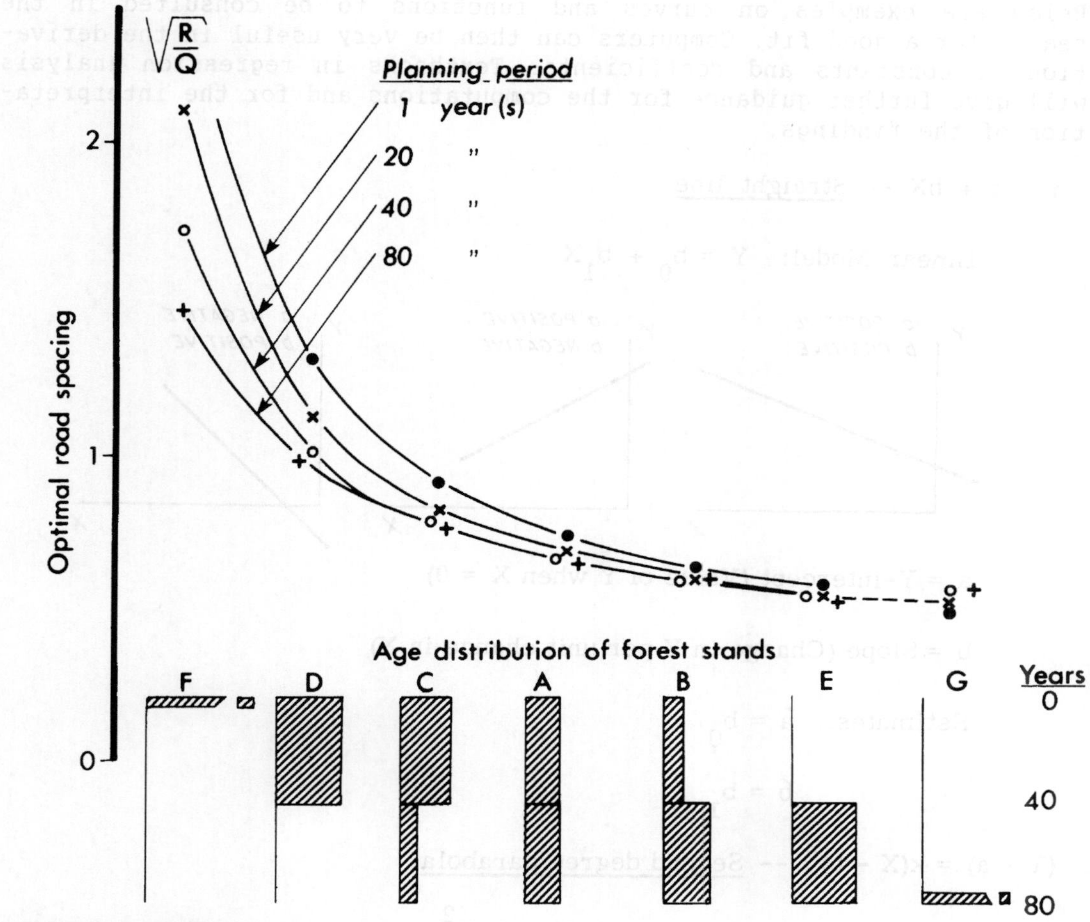

	F	D	C	A	B	E	G
Bare ground	100	–	–	–	–	–	–
0 – 40 y.	–	100	75	50	25	–	–
40 – 80 y.	–	–	25	50	75	100	–
Mature	–	–	–	–	–	–	100

Figure 3.1. Graph exemplifying the relative value of $\frac{R}{Q}$, which is proportional to the optimal road spacing. The computation is made for a predetermined management scheme for the forest stands, at different age distributions and at planning periods of 1, 20, 40 and 80 years. Rate of interest = 5 %. (See further Sundberg, 1963.)

B. SOME CURVE FORMS AND FUNCTIONS

The selection of a mathematical function for fitting relationships between the variables for a set of observations should in the first place be based on logics. The researcher shall use his insight and knowledge of the problem and establish a hypothesis for the relationship. This phase can be supported by trial plotting of the observations in a graph, often also grouping the observations in strata in order to explore the structure.

Below are examples on curves and functions to be consulted in the search for a good fit. Computers can then be very useful in the derivation of constants and coefficients. Textbooks in regression analysis will give further guidance for the computations and for the interpretation of the findings.

$Y = a + bX$ -- Straight line

Linear Model: $Y = b_0 + b_1X$

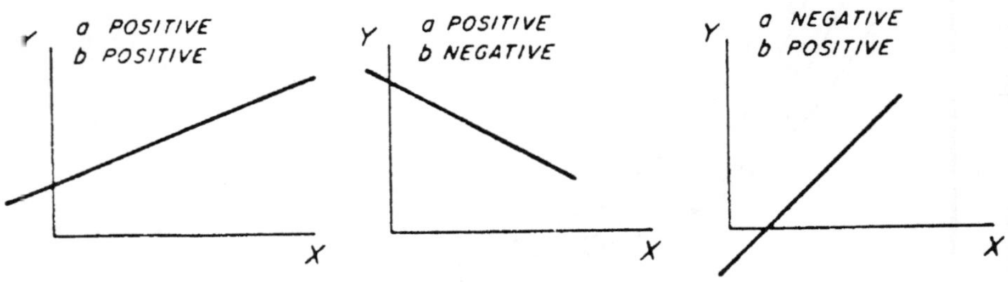

a = Y-intercept (Value of Y when X = 0)

b = Slope (Change in Y per unit change in X)

Estimates: $\hat{a} = b_0$

$\hat{b} = b_1$

$(Y - a) = k(X - b)^2$ -- Second degree parabola

Linear Model: $Y = b_0 + b_1X + b_2X^2$

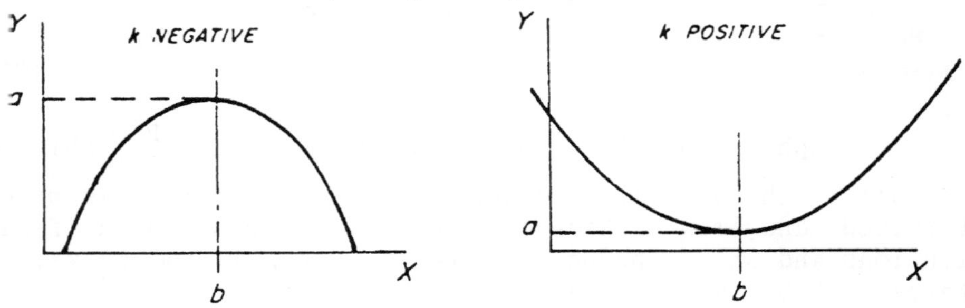

Y-intercept is at $Y = kb^2 + a$

X-intercepts are at $X = b \pm \sqrt{\frac{-a}{k}}$ (complex if $\frac{-a}{k}$ is negative)

Estimates: $\hat{a} = b_0 - (b_1^2/4b_2)$

$\hat{b} = -b_1/2b_2$

$\hat{k} = b_2$

$(Y - a) = k/X$ -- Hyperbola

Linear Model: $Y = b_0 + b_1 \left(\frac{1}{X}\right)$

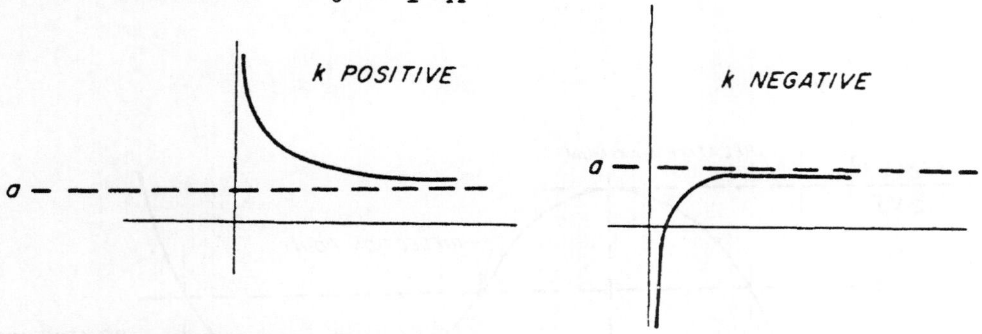

a = Level of horizontal asymptote

Estimates: $\hat{a} = b_0$

$\hat{k} = b_1$

$(Y - a) = \frac{k}{X} + bX$

Linear Model: $Y = b_0 + b_1 \left(\frac{1}{X}\right) + b_2 X$

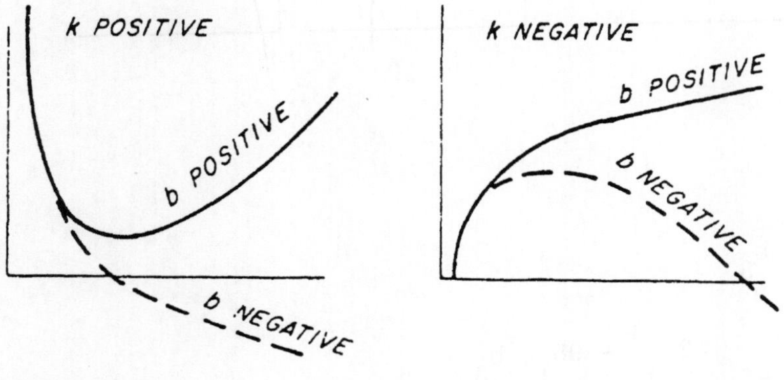

Estimates: $\hat{a} = b_0$, $\hat{k} = b_1$, $\hat{b} = b_2$

$$(Y - a) = k(X - b)(X - c)\left(X - \frac{b + c}{2}\right) \text{ -- } \underline{\text{Cubic}}$$

Linear Model: $Y = b_0 + b_1X + b_2X^2 + b_3X^3$

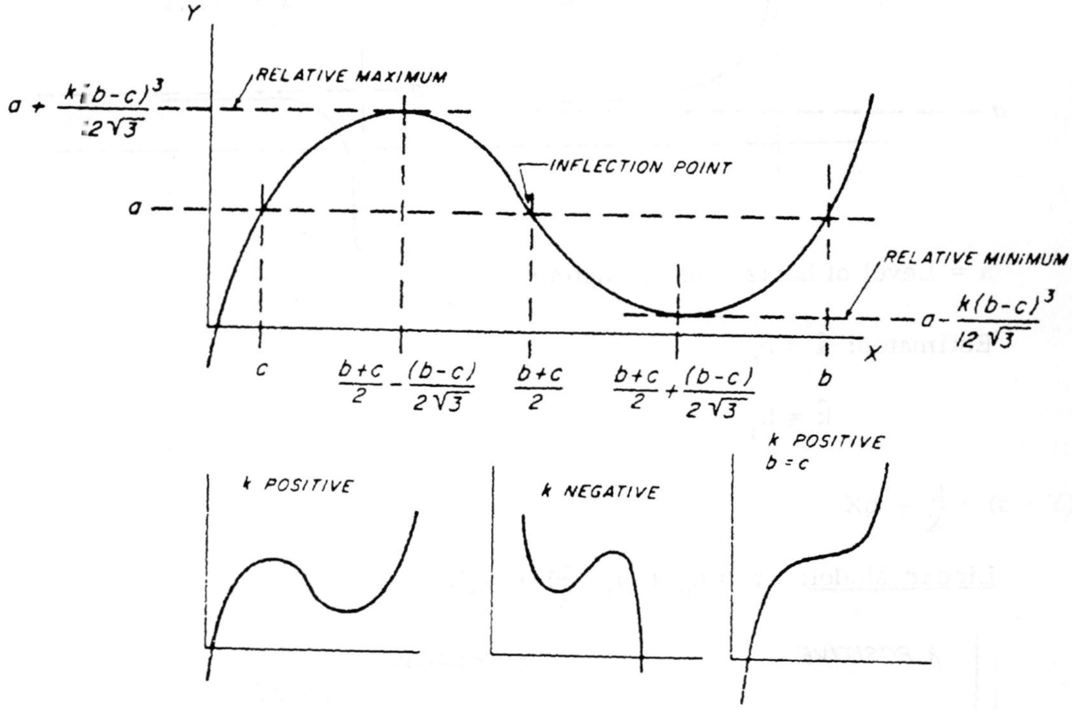

Estimates: $\hat{k} = b_3$

$$\hat{a} = b_0 + \frac{2b_2^3 - 9b_1b_2b_3}{27b_3^2}$$

$$\hat{b} = \frac{-b_2 + \sqrt{3(b_2^2 - 3b_1b_3)}}{3b_3}$$

$$\hat{c} = \frac{-b_2 - \sqrt{3(b_2^2 - 3b_1b_3)}}{3b_3}$$

$$Y = aX^b c^X$$

Linear Model: $\text{Log } Y = b_0 + b_1 \log X + b_2 X$

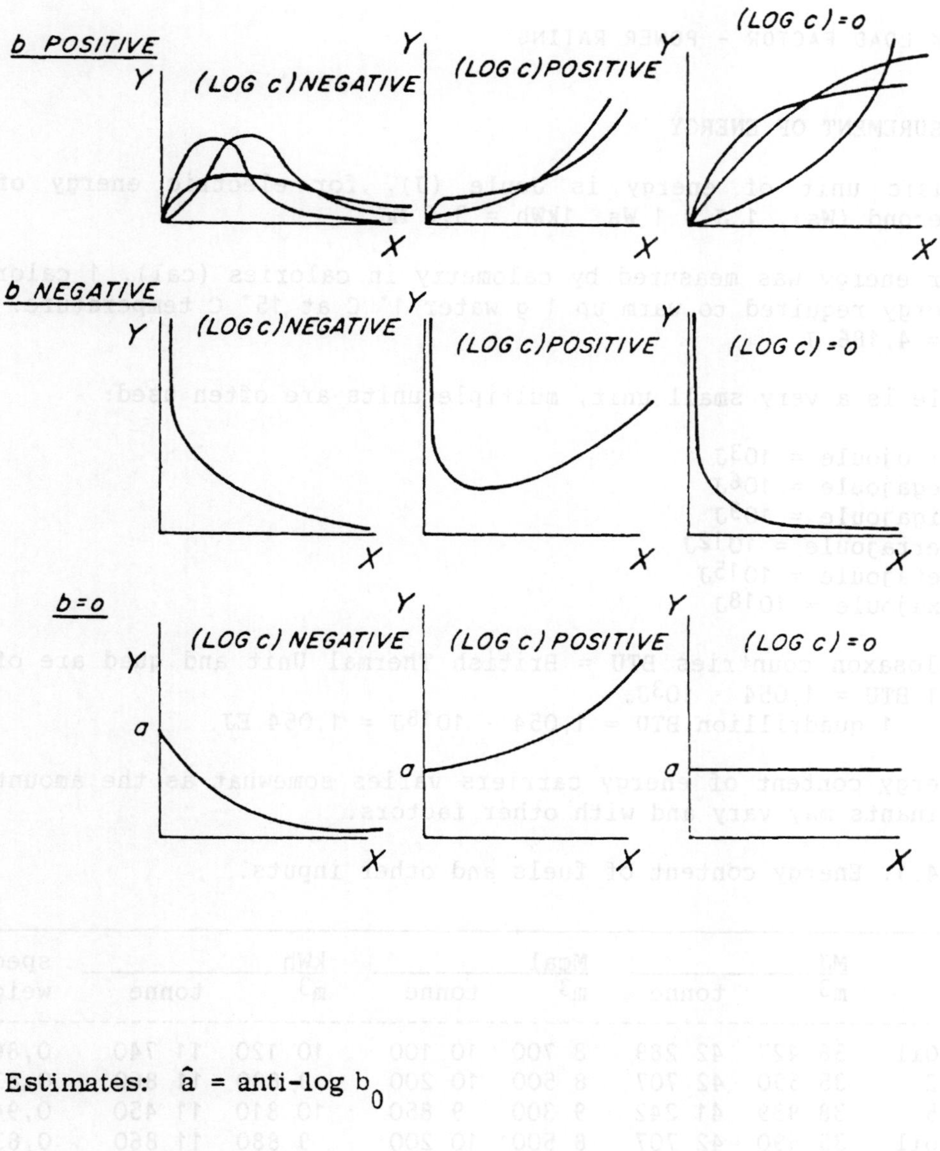

Estimates: $\hat{a}$ = anti-log b_0

$\hat{b} = b_1$

$\hat{c}$ = anti-log b_2

APPENDIX TO THE CHAPTER 4. ENERGY ANALYSIS

A. MEASUREMENT OF ENERGY

B. WOOD

C. WORK LOAD FACTOR - POWER RATING

A. MEASUREMENT OF ENERGY

The basic unit of energy is Joule (J), for electric energy often watt-second (Ws), 1 J = 1 Ws. 1kWh = 3,6 MJ.

Earlier energy was measured by calometry in calories (cal), 1 calory = the energy required to warm up 1 g water 1° C at 15° C temperature.
1 cal = 4,186 J.

As Joule is a very small unit, multiple units are often used:

kJ = kilojoule = 10^3J
MJ = megajoule = 10^6J
GJ = gigajoule = 10^9J
TJ = terrajoule = 10^{12}J
PJ = petajoule = 10^{15}J
EJ = exajoule = 10^{18}J

In anglosaxon countries BTU = British Thermal Unit and quad are often used, 1 BTU = 1,054 · 10^3J.
1 quad = 1 quadrillion BTU = 1,054 · 10^{18}J = 1,054 EJ.

The energy content of energy carriers varies somewhat as the amount of contaminants may vary and with other factors.

Table 4.1. Energy content of fuels and other inputs.

Fuels	MJ		Mcal		kWh		spec.
	m^3	tonne	m^3	tonne	m^3	tonne	weight
Crude Oil	36 427	42 289	8 700	10 100	10 120	11 740	0,86
Oil 1-2	35 590	42 707	8 500	10 200	9 880	11 860	0,83
Oil 3-5	38 939	41 242	9 300	9 850	10 810	11 450	0,94
Dieseloil	35 590	42 707	8 500	10 200	9 880	11 860	0,83
Kerosene	34 961	43 126	8 350	10 300	9 710	11 980	0,81
Benzin	31 403	43 126	7 500	10 300	8 720	11 980	0,73
Fossil coal	-	27 360	-	6 536	-	7 600	
Coke	-	28 053	-	6 700	-	7 790	
Gas	33 480	-	-	-	9 300	-	

Other inputs - typical values:

1 kg equipment	21 000 Kcal
1 kg nitrogen fertilizer	14 700 "
1 kg phosphorus fertilizer	3 000 "
1 kg potassium fertilizer	1 600 "
1 kg herbicides	57 000 "
1 kg fungicide	22 000 "
1 kg insecticide	44 000 "
1 kg irrigation water	2 "

B. WOOD

Wood often contains water. Also its chemical composition may vary somewhat from specie to specie but this variation is small and can usually be neglected.

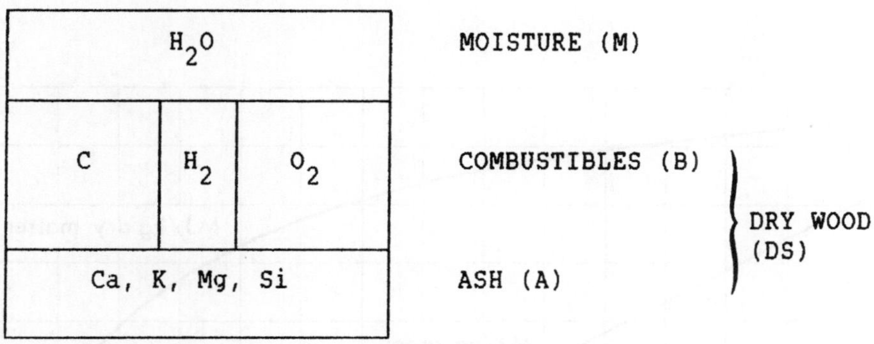

Figure 4.1. The composition of wood

The combustible fraction (B) is normally composed of:

0,515 C
0,060 H_2
0,425 O_2

Ashfree dry wood contains as an average 19,2 MJ/kg (M = 0, A = 0). The ash content is small, around 0,005.

The following formula may be used for determination of the energy value:

$$W_{eff} = 35,9 \cdot C \cdot B + 101,5 \cdot B(H - \frac{O}{8}) - \frac{2,45 \cdot M}{100 - M} \qquad (4.1)$$

W_{eff} = effective energy value, MJ/kg dry matter
C = fraction of carbon of B
H = fraction of hydrogen of B
O = fraction of oxygen of B
M = moisture content, in %
B = fraction of combustibles of dry material B = DS · (1-A)

For practical purpose the following applies. Assume that a wood user and a seller agrees on a price per net heat content of the wood. It is computed on the basis of tests of the moisture content of representative samples of the wood.

$$\text{Net heat content} = W_{eff} \cdot \frac{D_t}{100} - (2,45 \cdot \frac{100-D_t}{100}) \text{ MG kg}^{-1} \quad (4.2)$$

Effective heat content, W_{eff} 19,2 MJ kg^{-1} DS
DS = dry substance
Vaporization heat of water at 25° C = 2,45 MJ kg^{-1}

<u>Example</u>: dry matter 55%, moisture 45%

$$\text{Net heat } 19,2 \cdot \frac{55}{100} - 2,45 \cdot \frac{45}{100} = 9,5 \text{ MJ kg}^{-1} \text{ green weight}$$

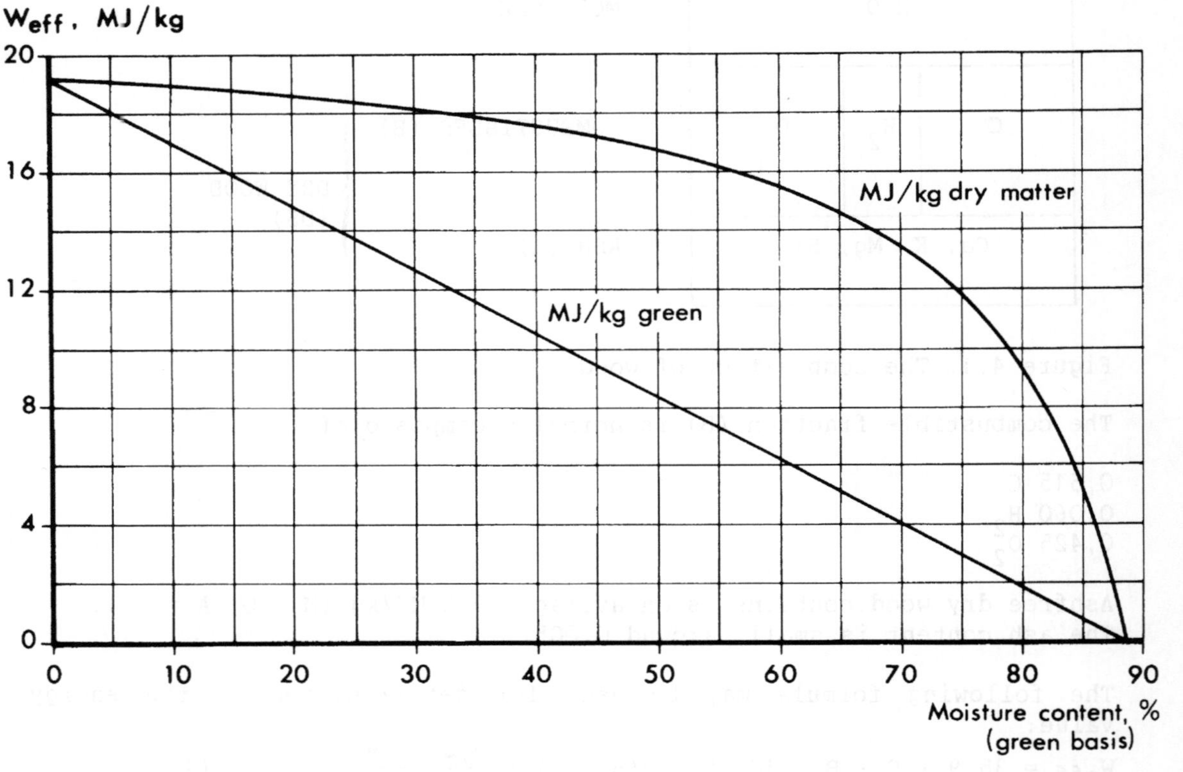

Figure 4.2. The calorific energy value of wood at different moisture content. MJ per Kg dry matter resp. green wood. (Dehlén, Mattsson, 1980)

It should be noted that for moist wood, its energy value compared to oil or coal, is often further reduced during combustion in ovens as the efficiency of the oven decreases with increasing moisture in the wood.

Dehlén and Mattsson (1980) gives the following example on the equivalents of wood and fuel oil taking into account their respective efficiencies in oven combustion.

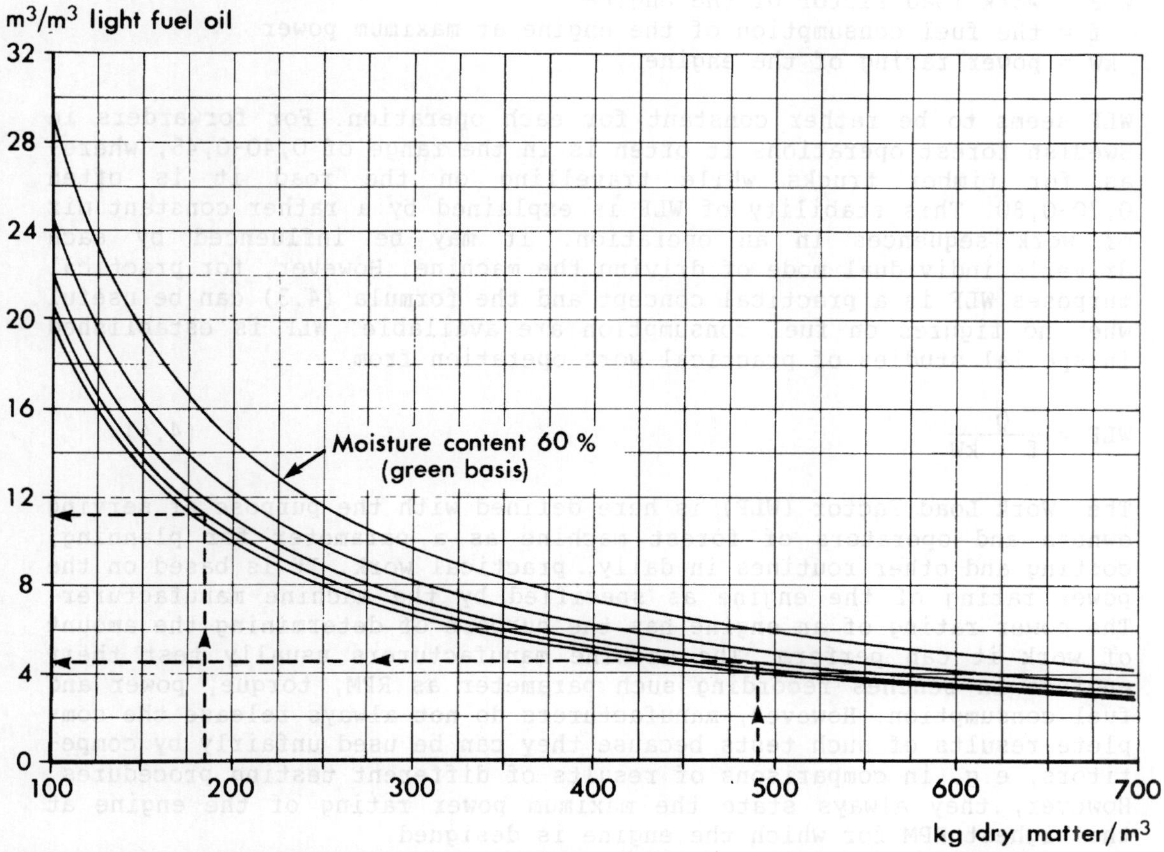

Figure 4.3. Wood quantity, in Kg dry matter to replace 1 cubic meter oil as function of moisture content, top curve 60%, next curves 50, 40, 30, 20 successively. Oven efficiency for oil 0,9 for wood 0,8-0,9.

Softwood normally 400-500 Kg/m^3 dry matter
Hardwoods normally 550-650 Kg/m^3 dry matter, more for some tropical woods

C. WORK LOAD FACTOR - POWER RATING

The kW-power rating (or HP) of a machine is as a rule given in the machine specification from the manufacturer and also often the engine type. The energetic efficiency of each engine type is rather well established, sometimes it is also stated in the engine specification or by official machine testing station. The fuel consumption of a machine in practical work can then be derived from the expression:

$$Q = WLF \cdot f \cdot kW \tag{4.3}$$

where

Q = fuel consumption per unit time
WLF = work load factor of the engine
f = the fuel consumption of the engine at maximum power
kW = power rating of the engine

WLF seems to be rather constant for each operation. For forwarders in Swedish forest operations it often is in the range of 0,40-0,45, whereas for timber trucks while travelling on the road it is often 0,70-0,80. This stability of WLF is explained by a rather constant mix of work sequences in an operation. It may be influenced by each driver's individual mode of driving the machine. However, for practical purposes WLF is a practical concept and the formula (4.3) can be useful when no figures on fuel consumption are available. WLF is established in special studies of practical work operation from

$$WLF = \frac{Q}{f \cdot kW} \tag{4.4}$$

The "Work Load Factor (WLF) is here defined with the purpose of serving owners and operators of forest machine as a parameter for planning, costing and other routines in daily, practical work. It is based on the power rating of the engine as specified by the machine manufacturer. The power rating of an engine has the purpose of determining the amount of work it can perform. The machine manufacturers usually test their engines in benches recording such parameter as RPM, torque, power and fuel consumption. However, manufacturers do not always release the complete results of such tests because they can be used unfairly by competitors, e.g. in comparisons of results of different testing procedures. However, they always state the maximum power rating of the engine at the highest RPM for which the engine is designed.

Engine power ratings are not always comparable. In the metric system 1 HP (DIN) = 75 KGM/s where as 1 HP (ASA) = 33 000 foot pound per minute (difference 1,5%).

Test protocols are established for net rating by DIN and for net and gross rating by ISO, SAE and BSAU. There are some minor difference with regard to temperature and air pressure but the main differences are regarding auxilary engine equipment, summed up below:

1. In all tests the hydraulic pump (if any) is decoupled.
2. In the DIN test the generator and air compressor are unloaded.
3. In comparison with the DIN protocol the other three remove the following parts:

	Gross	Net
ISO	Air intake, muffler, fan, generator, air compressor	Air compressor
SAE	Muffler, fan, generator, air compressor	Air compressor
BSAU	Fan, Generator, air compressor	

Note: ISO = International Standardisation Organization
DIN = Deutsche Industrienormen (West Germany)
SAE = Society of Automotive Engineers (USA)
BSAU = British Standard Automobile

In the OECD countries a special protocol is established for the testing of farm tractor based on the DIN protocol but braking the engine on the fully equiped tractor's power take off instead of the flywheel, the difference being in the order of 3-5%. In USA farm tractors are often tested according to the Nebraska protocol, which is almost identical with the OECD protocol.

The power rating on an engine fully stripped on auxilary equipment can result in a gross power rating up to 35% higher than when measured on the power take off according to the OECD protocol for farm tractors. The figure (4.4) illustrate the case.

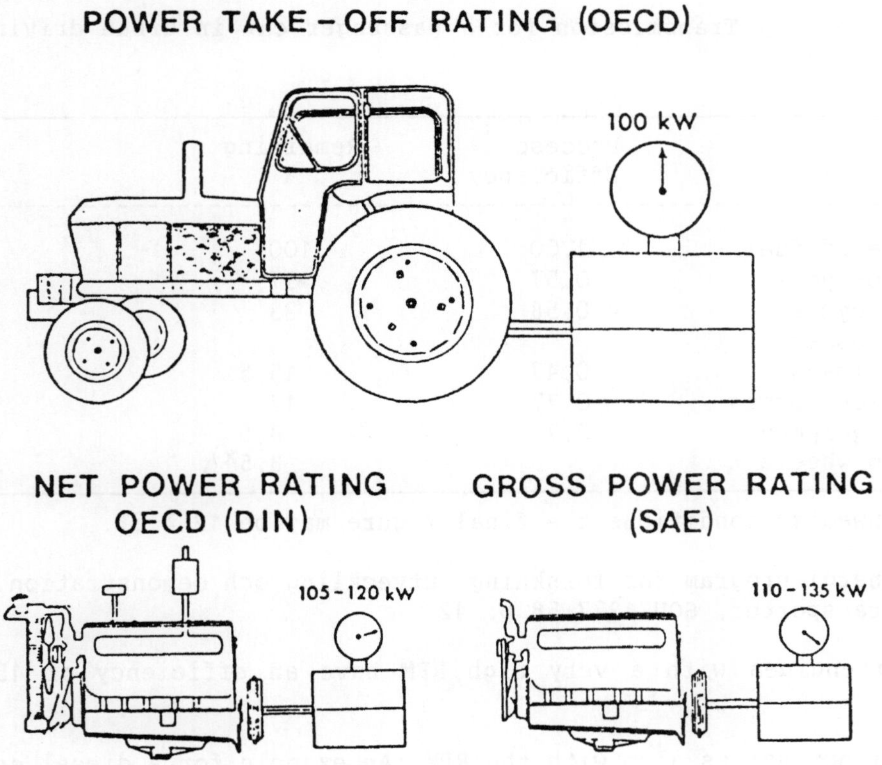

Figure 4.4. Comparison of power ratings.

Thus the numeraires on power rating of an engine depend on where on the power transmission the measurement is made and what auxilary equipment attached to the engine that actually was powered by it during the test.

The efficiency of an engine is defined as the ratio between the energy (or heat) content of the fuel to the work performed. It can be illustrated by the following diagram for an efficiency of 35%.

1 liter Diesel Oil	->	Chem. Energy 9,880 kWh	->	Thermal Energy	->	η = 35%	->	Work 3,458 kWh

or

1:3,458 = 0,289 liter diesel oil per kW engine rating.

Thus this engine has transformed one litre of diesel oil to 3,458 kWh which can be used for actual work. 65% or 6,422 kWh dissappeared as heat.

The different designs of a diesel engine result in various efficiencies. A large marine engine with a low RPM may achieve around 50% efficiency. The most modern truck engines may go up to 42-43%. Tractor engines now in use seldom exceed 35-36%, older types may often be down to about 30%.

An example of a benzine engine is shown below:

Table 4.2. Energy Transmission for a passenger car in urban driving in USA

	Process Efficiency	Remaining %
Heat value of fuel	1,00	100
Ideal Otto-cycle	0,57	57
Real Otto-cycle	0,58	33
Friction losses partial loads	0,47	15,5
Transmission losses	0,77	12
Auxilary equipment	0,71	8,5
Remains to wheels		8,5x)

x) Under Swedish conditions the final figure may be 12 ± 3%.

Source: Energi-program för forskning, utveckling och demonstration. Transporter, SOU 1977:58 p. 42.

Chain saw engines with a very high RPM have an efficiency of 15-20% only.

The efficiency varies also with the RPM. An example for a diesel engine is shown in the mussel diagram, figure 1. In this case, the optimum efficiency is for RPM = 1500 and decreases with around 10% at the RPM = 2200 at which the (maximum) power rating of the engine occurs and is measured.

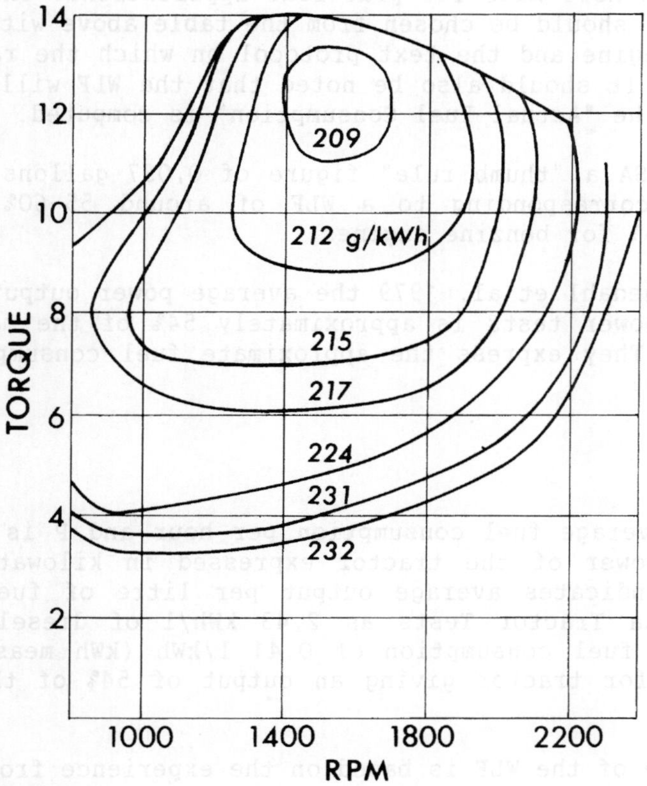

Figure 4.5. Mussel diagram for a diesel engine (example)

The following table shows the approximative fuel consumption per kWh for different engines at their maximum power rating, assuming an increase in fuel consumption of 5-10% above optimum efficiency.

Table 4.3. Fuel consumption per kWh at maximum power rating

	Otto engines		Diesel engines		
	chain saws	cars	last design	modern types	early 70's
(at max power)	17%	25%	38%	33%	28%
Fuel consumption					
liter/kWh	0,67	0,46	0,27	0,31	0,36
g/kWh	0,49	0,33	0,22	0,26	0,30

<u>Note</u>: 1 lit diesel oil = 9,88 kWh. Sp. weight = 0,83
1 lit benzine = 8,72 kWh. Sp. weight = 0,73

The work load factor (WLF) is then:

$$WLF = \frac{\text{Actual Fuel Consumption per Hour at Work}}{\text{Power Rating of Engine in kW} \cdot \text{Fuel Consumption per kWh}}$$

The definition is here made for practical applications. The "Fuel Consumption per kWh" should be chosen from the table above with due regard to the type of engine and the text protocol on which the rating of the engine is based. It should also be noted that the WLF will depend over which unit time the "Actual Fuel Consumption" is computed.

In forestry in USA a "thumb rule" figure of 0,037 gallons per HP-hour is often used, corresponding to a WLF of around 55-60% for diesel engines and 40-45% for benzine engines.

According to Liljedahl et al. 1979 the average power output of a tractor in various power tests is approximately 54% of the maximum power take off power. They express the approximate fuel consumption in the following formula:

$$Q = \frac{0{,}54 \cdot P}{kWh/l} \qquad (4.5)$$

Where Q is the average fuel consumption per hour and P is the maximum power take-off power of the tractor expressed in kilowatts. The expression kWh/l indicates average output per litre of fuel, which is given by Nebraska Tractor Tests as 2,43 kWh/l of diesel fuel. This corresponds to a fuel consumption of 0.41 l/kWh (kWh measured at the power take-off) for tractor giving an output of 54% of their maximum rating.

The practical use of the WLF is based on the experience from the rather scanty measurements of fuel consumption in practical work made up to now that the WLF is rather constant for each type of operation. If confirmed, the WLF can then be used as a control parameter for each machine or driver. It can also be used to forecast the fuel consumption for different machines on the basis of which the total machine cost may be assessed, as outlined in the main text of this report (chapter 4.8.3, p. 146

REFERENCES

Abegg, B. (1978). Die Schätzung der optimale Dichte von Waldstrassen im Traktorfaturbaren Gelände. Mitteilungen/Eidgenössiche Anstalt für das Forstliche Versuchswesen 54 (2).

American Pulpwood Association (1975). Fuel Survey. Washington, D.C.: A.P.A.

Ash, M.J., Knoblock, P.C. & Peters, N. (1980). Energy Analysis of Energy from the Forest Options. ENFOR Project P-59, Canadian Forestry Service.

Åstrand, P.-O. & Rodahl, K. (1970). Textbook of work physiology. New York.

Axelson, O. (1974). Heat Stress in Forest Work. Rome: FAO.

Backmund, F. (1966). Kennzahlen für den Grad der Erschliessung von Forstbetrieben durch Autofahrbare Wege. Forstwissenschaftliches Centralblatt 85 (12), 342-354.

Baltic, T.J. & Betters D.R. (1983). Net Energy Analysis of a Fuelwood Energy System. Resources and Energy (5), 45-64.

Bendz, M. (1974). Logging and Log Transport in Tropical High Forest. FAO Forestry Development Paper, No. 18. Rome.

Bjerkelund, T.C. (1979). The Long-Distance-Off and On-Road Transportation Concept for Tree Harvesting. (2 ed.) Fredericton: UNB.

Christensen, E.H. (1955). Physical Working Capacity of old workers and physiological background for work tests and work evaluations. World Health Org., (13), 587.

Clark, C. & Haswell, M. (1970). The Economics of Subsistence Agriculture. Macmillan and Co Ltd, London, 4th ed.

Cleveland, C.J., Costanza, R., Hall, C.A.S. & Kaufmann, R.K. (1984). Energy and the U.S. Economy: A biophysical perspective. Science (225), 890-897.

Cobb, C.W. & Douglas, P.H. (1928). A Theory of Production. The American Economic Review, Supplement 18.

Costanza, R. & Heredeen, R.A. (1984). Embodied energy and economic value in the U.S. economy: 1963, 1967 and 1972. Resources and Energy 6:129.

Costanza, R. (1980). Embodied energy and economic valuation. Science 210, 1219-1224.

Cubbage, F. (1981). Machine Rate Calculations and Productivity Rate Tables of Harvesting Southern Pine. Staff paper series 24./USDA Southern Forest Experiment Station, New Orleans.

Daly, H.E. (1977). Steady State Economics. W.H. Freeman.

Daly, H.E. (1983). The circular flow of exchange value and the linear throughput of matter energy: A case of misplaced concreteness. Faculty Working Paper, Ec. Dept of Lousianna State University, Baton Rouge.

Dehlen, R. & Mattsson, J.E. (1980). Vad är skogsenergin värd: beräkningsmetodik och omföringstal. Information från Projekt Skogsenergi, 10. Garpenberg.

Drucker, P.F. (1982). Towards the next economics. In: The Crisis in Economic Theory. D Bell and J Kristo (eds) New York: Basic Books.

Duerr, William A. (1960). Fundamentals of Forestry Economics. New York.

Erickson, John R. (1983). The Role of Biomass as an Energy Source in the United States. USDA-Forest Service.

FAO/ECE (1985). Outlook for the use of wood for energy. TIM/EFC/AC.3/ R.22 10 July, 1985. Geneva.

Fontaine, R.G. (1971). The forestry sector in today's setting. Unasylva, Volume 25 (1), No 100, p. 9-14.

Gasslander, J.E., Mattsson J.E. & Sundberg U. (1979). Study on the Energy Requirements in Forest Logging. Sw Univ of Agricultural Sciences, Dept of Operational Efficiency, Garpenberg.

Georgescu-Roegon, N. (1971). The Entropy Law and the Economic Process. Cambridge, Harvard Univ. Press.

Georgescu-Roegon, N. (1975). Energy and Economic Myths. Southern Econ. J. 41: 347-381.

Georgescu-Roegon, N. (1976). Energy and Economic Myths. Pergamon Press.

Georgescu-Roegon, N. (1979). Energy analysis and economic evaluation. Southern Econ. J. 45: 1023-1058.

Golob, Theodore B. (1986). Analysis of Short Rotation Forest Operations. Ottawa. Nat. Res. Council, Canada. NCCC No. 26014.

Guimier, D.Y. & Wellburn, G.V. (1984). Logging with heavy-lift airships. Vancouver: FERIC. (Technical report Forest engineering research institute of Canada. No TR-58).

Hall, C.A.S., Cleveland, C.J. & Kaufmann, R. (1986). Energy and Resource Quality. The Ecology of the Economic Process. New York: John Wiley & Sons.

Hall, C.A.S. & Day, J. (1977). Ecosystem Modelling in Theory and Practice. New York: Wiley - Interscience.

Hall, D.O., Myers, N. & Margaris, N.S. (1985). Economics of ecosystems management. Dordrecht: Junk Publishers.

Inman, Robert J. An Energy Budget for Hypothetical Biomass Plantation. (Stencil, Stanford Res. Inst.).

IFIAS (1974). Workshop on energy analysis, methodology and conventions. (International Federation of Institutes for Advanced Study). Report 6, Solna.

King, Hubert M. (1971). The Energy Resources of the Earth. (Energy and Power, 31). San Francisco: W.H. Freeman and Co.

Koroleff, A. (1961). New Approach to Collective Work Evaluation of Machinery Aggregation. Pulp Paper Mag. Canada.

Larsson, G. (1959). Studies on Forest Road Planning. Transactions No. 147. Royal Inst. Techn. Stockholm.

Larsson, G. & Rydstern, O. (1968). Economic design of motor truck road systems in forest areas. Acta Polytechnica Scandinavia Ci 52.

Leach, Gerald (1976). Energy and Food Production. Guilford. IPC Press Ltd.

Leontief, W.W. (1982). Academic economics. Science 217: 104-107.

Liljedahl, J.B., Carletonne, W.M., Turnquist, P.K. & Smith, D.W. (1979). Tractors and their power units. New York: John Wiley and Sons.

Ministerium f. Ernährung, Landwirtschaft, Umwelt und Forsten: Jahresbericht der Landesforstveraltung 1984.

Odum, H.T. & E.C. (1976). Energy Basis for Man and Nature. New York: McGraw Hill.

Oseni, A.M. & Ward, D.M. (1972). Problems of Forest Ergonomics in Tropical Countries. Nigeria Forestry Bulletin No. 21, Fed. Dept of Forestry. Ibadan.

Peters, Penn A. (1978) Spacing of Roads and Landings to Minimize Timber Harvest Cost. Forest Sci., Vol. 24, No 2 pp 209-217.

Pimentel, D. & Hall, C.W. (1984). Food and Energy Resources. Orlando: Academic Press, Inc.

Prins, Kit (1979). Energ derived from wood in Europe, the USSR and North America. Unasylva, vol 31, 123.

Reaven, Sheldon J. (1986). Perspectives on convergence in net energy analysis. Washington, D.C.: In Energy from Biomass and Wastes.

Rehnborg, U. & Uhlin, H.E. (1975). Skogs- och jordbruket i energikris. Energiflödet i svenskt skogs- och jordbruk 1956 och 1972. Skogs- och lantbruksakademins Tidskrift 114.

Richardson, S.D. (1967). Manpower and training requirements in forestry development planning. Paper to the 3rd session of the FAO Advisory Committee on Forestry Education. FAO. Rome.

Rumpf, Janos (1986). Ein Energiebezogenes Modell der Holzernte und Matrialbringung. Univ. Forstwirtschaft und Holzind. Sopron, Hungary.

Samset, Ivar (1985). Winch and Cable Systems. Dordrecht: Martinus Nijhoff/Dr. W. Junk Publishers.

Schurr, S.H. & Netschert, B. (1980). Energy in the American Economy. 1850-1975. Baltimore: John Hopkins Univ. Press.

Segebaden, G. von (1964). Studies of Cross-Country Transport Distances and Road Net Extensions. Studia Forestalia Suecia 18.

Segebaden, G. von (1969). Studies on the Accessibility of Forest and Forest Land in Sweden. Studia Forestalia Suecia.

Slesser, M. (1978). Energy in the Economy. New York: St Martinus Press.

Smil, V. & Kuz, T. (1976) European Energy Elasticities. Energy Policy vol. 4.

Sundberg, U. (1963). Some views on the theory of planning a forest road network in non-alpine regions. Log/symposium 1963-01-16. FAO/EFC, Geneva.

Sundberg, U. (1965). Distribution of the costs of joint forest roads according to crosswise and lengthwise road functions. Studia Forestalia No 30, Garpenberg.

Sundberg, U. (1973). An example on estimating logging transport and road costs in an afforestation venture. (Lecture, FAO/SIDA Forest Inventory Course) Rome: FAO.

Sundberg, U. (1979). An Analysis of Mechanization in Forestry - a Method Study. (Report. Swedish University of Agricultural Sciences. Dept of Operational Efficiency, no 129) Garpenberg: SLU.

Sundberg, U. (1982). A Study on Cost of Machine Use in Forestry - Proposing Fuel Consumption as Cost Determinant. (Report. Swedish University of Agricultural Sciences. Dept of Operational Efficiency, no 142) Garpenberg: SLU.

Sundberg, U. (1984). The Charcoal Burner at Basttjärn. In "Forest Energy in Sweden", The Forestry Energy Project, Swedish University of Agricultural Sciences, Stencil no 281, Garpenberg: SLU.

Sundberg, U. & Svanqvist, N. (1987). Fuel consumption as indicator of the economics in mechanization. Scandinavian Journal of Forest Research 2:1987.

Svanqvist, N. (1985) Fuel Consumption as Cost Indicator. (Report. Swedish University of Agricultural Sciences. Dept of Operational Efficiency, Uppsatser och Resultat no 11) Garpenberg: SLU.

UN/ECE (1982). Papers presented to the seminar on energy conservation and self-sufficiency in the sawmilling industry. Bonn.

Worrel, A.C. (1959). Economics of American Forestry. New York: John Wiley & Sons.

Zerbe, J.I., Sundberg, U. & Erickson, J.R. (1986). Energy from the Forest as a Contributor to and Measure of Economic Growth. Paper for the 18th IUFRO World Congress Sept, 1986, in Ljubljana Yugoslavia.

UNITS AND CONVERSION FACTORS

The International System of Units, SI (selected list)

Base Units

length	metre	m
mass	kilogram	kg
time	second	s

Derived Units

force	newton	N	$1N = 1kg \cdot m/s^2$
energy	joule	J	$1J = 1N \cdot m$
power	watt	W	$1W = 1J/s$

Multiples

10^{18}	exa	E	10^{-1}	deci	d
10^{15}	peta	P	10^{-2}	centi	c
10^{12}	tera	T	10^{-3}	milli	m
10^{9}	giga	G	10^{-6}	micro	µ
10^{6}	mega	M	10^{-9}	nano	n
10^{3}	kilo	K	10^{-12}	pico	p
10^{2}	hecto	h	10^{-15}	femto	f
10	deca	da	10^{-18}	atto	a

Normally, choose the prefix so that the numerical value falls between 0,1 and 1000. Prefered prefixes underlined.

Additional Units

degree	°	1° = 1/360 of a full circle
litre	l	1 l = 1 dm^3
minute (time)	min	1 min = 60 s
hour	h	1 h = 3600 s
day	d	1 d = 24 h
tonne	t	1 t = 1000 kg

Length

m	in (inch)	ft (foot)	yd (yard)	mile	nautical mile
1	39,370	3,280 8	1,093 6	0,621 37 · 10^{-3}	0,539 96 · 10^{-3}
25,4 · 10^{-3}	1	83,333 · 10^{-3}	27,778 · 10^{-3}	15,783 · 10^{-6}	13,715 · 10^{-6}
0,304 8	12	1	0,333 33	0,189 39 · 10^{-3}	0,164 58 · 10^{-3}
0,914 4	36	3	1	0,568 18 · 10^{-3}	0,493 74 · 10^{-3}
1,609 3 · 10^{3}	63,36 · 10^{3}	5,28 · 10^{3}	1,76 · 10^{3}	1	0,868 98
1,852 · 10^{3}	72,913 · 10^{3}	6,076 1 · 10^{3}	2,025 4 · 10^{3}	1,150 8	1

1Å (1 ångström) = 10^{-10} m

Area

m^2	in^2	ft^2	yd^2	acre	square mile
1	1,550 0 · 10^{3}	10,764	1,196 0	0,247 10 · 10^{-3}	0,386 10 · 10^{-6}
0,645 16 · 10^{-3}	1	6,944 4 · 10^{-3}	0,771 61 · 10^{-3}	0,159 42 · 10^{-6}	0,249 10 · 10^{-9}
92,903 · 10^{-3}	144	1	0,111 11	22,957 · 10^{-6}	35,870 · 10^{-9}
0,836 13	1,296 · 10^{3}	9	1	0,206 61 · 10^{-3}	0,322 83 · 10^{-6}
4,046 9 · 10^{3}	6,272 6 · 10^{6}	43,56 · 10^{3}	4,84 · 10^{3}	1	1,562 5 · 10^{-3}
2,590 0 · 10^{6}	4,014 5 · 10^{9}	27,878 · 10^{6}	3,097 6 · 10^{6}	640	1

1 ha (1 hectare) = 10 000 m^2

Volume

m^3	in^3	ft^3	yd^3	UK gallon	US gallon
1	61 024 · 10^{3}	35,315	1,308 0	219,97	264,17
16,387 · 10^{-6}	1	0,578 70 · 10^{-3}	21,434 · 10^{-6}	3,604 6 · 10^{-3}	4,329 0 · 10^{-3}
28,317 · 10^{-3}	1,728 · 10^{3}	1	37,037 · 10^{-3}	6,228 8	7,480 5
0,764 56	46,656 · 10^{3}	27	1	168,18	201,97
4,546 1 · 10^{-3}	277,42	0,160 54	5,946 1 · 10^{-3}	1	1,201 0
3,785 4 · 10^{-3}	231	0,133 68	4,951 1 · 10^{-3}	0,832 68	1

1 l (1 litre) = 1 dm^3 = 10^{-3} m^3

Velocity

m/s	km/h	ft/s	mile/h	kn (knot)
1	3,6	3,280 8	2,236 9	1,943 8
0,277 78	1	0,911 34	0,621 37	0,539 96
0,304 8	1,097 3	1	0 681 82	0,592 48
0,447 04	1,609 3	1,466 7	1	0,868 98
0,514 44	1,852	1,687 8	1,150 8	1

1 kn = 1 nautical mile per hour

Density

kg/m^3	g/cm^3	lb/in^3	lb/ft^3
1	10^{-3}	36,127 · 10^{-6}	62,428 · 10^{-3}
10^{3}	1	36,127 · 10^{-3}	62,428
27,680 · 10^{3}	27,680	1	1,728 · 10^{3}
16,019	16,019 · 10^{-3}	0,578 70 · 10^{-3}	1

Mass

kg	lb (pound)	slug	oz (ounce)	(long) cwt (hundred-weight)	(long) ton (UK)	sh cwt (short hundred-weight) (US)	sh tn (short ton) (US)
1	2,204 6	68,522 · 10^{-3}	35,274	19,684 · 10^{-3}	0,984 21 · 10^{-3}	22,046 · 10^{-3}	1,102 3 · 10^{-3}
0,453 59	1	31,081 · 10^{-3}	16	8,928 6 · 10^{-3}	0,446 43 · 10^{-3}	10 · 10^{-3}	0,5 · 10^{-3}
14,594	32,174	1	514,79	0,287 27	14,363 · 10^{-3}	0,321 74	16,087 · 10^{-3}
28,350 · 10^{-3}	62,5 · 10^{-3}	1,942 6 · 10^{-3}	1	0,558 04 · 10^{-3}	27,902 · 10^{-6}	0,625 · 10^{-3}	31,25 · 10^{-6}
50,802	112	3,481 1	1,792 · 10^{3}	1	50 · 10^{-3}	1,12	56 · 10^{-3}
1,016 1 · 10^{3}	2,24 · 10^{3}	69,621	35,84 · 10^{3}	20	1	22,4	1,12
45,359	100	3,108 1	1,6 · 10^{3}	0,892 86	44,643 · 10^{-3}	1	50 · 10^{-3}
907,19	2 · 10^{3}	62,162	32 · 10^{3}	17,857	0,892 86	20	1

In the USA 1 ounce is also called 1 avoirdupois ounce to distinguish it from 1 troy ounce (oz t)
7 680 oz avdp = 7 000 oz t

1 hundredweight (UK) is called 1 long hundredweight in the USA
1 ton (UK) is called 1 long ton in the USA

1 tonne ("metric ton") = 1 000 kg

Force

N	dyn	kgf (kp) (kilogram-force)	lbf (pound-force)
1	0,1 · 10^{6}	0,101 97	0,224 81
10 · 10^{-6}	1	1,019 7 · 10^{-6}	2,248 1 · 10^{-6}
9,806 6	0,980 66 · 10^{6}	1	2,204 6
4,448 2	0,444 82 · 10^{6}	0,453 59	1

Moment of force

Nm	kgf · m	lbf · in	lbf · ft
1	0,101 97	8,850 8	0,737 56
9,806 6	1	86,796	7,233 0
0,112 99	11,521 · 10^{-3}	1	83,333 · 10^{-3}
1,355 8	0,138 26	12	1

Energy

J Nm, Ws	kWh	kgf · m	kcal	metric horsepower-hour	ft · lbf (foot pound-force)	Btu (British thermal unit)
1	0,277 78 · 10^{-6}	0,101 97	0,238 85 · 10^{-3}	0,377 67 · 10^{-6}	0,737 56	0,947 82 · 10^{-3}
3,6 · 10^{6}	1	0,367 10 · 10^{6}	859,85	1,359 6	2,655 2 · 10^{6}	3,412 1 · 10^{3}
9,806 6	2,724 1 · 10^{-6}	1	2,342 3 · 10^{-3}	3,703 7 · 10^{-6}	7,233 0	9,294 9 · 10^{-3}
4,186 8 · 10^{3}	1,163 · 10^{-3}	426,94	1	1,581 2 · 10^{-3}	3,088 0 · 10^{3}	3,968 3
2,647 8 · 10^{6}	0,735 50	0,27 · 10^{6}	632,42	1	1,952 9 · 10^{6}	2,509 6 · 10^{3}
1,355 8	0,376 62 · 10^{-6}	0,138 26	0,323 83 · 10^{-3}	0,512 06 · 10^{-6}	1	1,285 1 · 10^{-3}
1,055 1 · 10^{3}	0,293 07 · 10^{-3}	107,59	0,252 00	0,398 47 · 10^{-3}	778,17	1

1 erg = 0,1 · 10^{-6} J

Power

W Nm/s, J/s	kgf · m/s	kcal/s	kcal/h	metric horsepower	horsepower (foot-pound system)	ft · lbf/s	Btu/h
1	0,101 97	0,238 85 · 10^{-3}	0,859 85	1,359 6 · 10^{-3}	1,341 0 · 10^{-3}	0,737 56	3,412 1
9,806 6	1	2,342 3 · 10^{-3}	8,432 2	13,333 · 10^{-3}	13,151 · 10^{-3}	7,233 0	33,462
4,186 8 · 10^{3}	426,94	1	3,6 · 10^{3}	5,692 5	5,614 6	3,088 0 · 10^{3}	14,286 · 10^{3}
1,163	0,118 59	0,277 78 · 10^{-3}	1	1,581 2 · 10^{-3}	1,559 6 · 10^{-3}	0,857 79	3,968 3
735,50	75	0,175 67	632,42	1	0,986 32	542,48	2,509 6 · 10^{3}
745,70	76,040	0,178 11	641,19	1,013 9	1	550	2,544 4 · 10^{3}
1,355 8	0,138 26	0,323 83 · 10^{-3}	1,165 8	1,843 4 · 10^{-3}	1,818 2 · 10^{-3}	1	4,626 2
0,293 07	29,885 · 10^{-3}	69,999 · 10^{-6}	0,252 00	0,398 47 · 10^{-3}	0,393 02 · 10^{-3}	0,216 16	1

Pressure, stress

Pa	bar	kgf/cm² at	kgf/mm²	torr (≈ mm Hg)	atm	lbf/in² (psi)
1	10 10^{-6}	10,197 10^{-6}	0,101 97 · 10^{6}	7,500 6 · 10^{-3}	9,869 2 · 10^{-6}	0,145 04 · 10^{-3}
100 · 10^{3}	1	1,019 7	10,197 · 10^{-3}	750,06	0,986 92	14,504
98,066 10^{3}	0,980 66	1	10 · 10^{-3}	735,56	0,967 84	14,223
9,806 6 · 10^{6}	98,066	100	1	73,556 · 10^{3}	96,784	1,422 3 · 10^{3}
133,32	1,333 2 · 10^{-3}	1,359 5 · 10^{3}	13,595 · 10^{-6}	1	1,315 8 · 10^{-3}	19,337 · 10^{-3}
101,32 · 10^{3}	1,013 2	1,033 2	10,332 · 10^{-3}	760	1	14,696
6,894 8 10^{3}	68,948 · 10^{-3}	70,307 · 10^{-3}	0,703 07 · 10^{3}	51,715	68,046 · 10^{-3}	1

1 mm water column ≈ 9,81 Pa 1 inch water column ≈ 249,09 Pa; 1 inch Hg ≈ 3 386,4 Pa

Temperature

Quantity	Kelvin scale	Celsius scale	Rankine scale	Fahrenheit scale	Physical characteristics
Temperatures associated together	0 K	−273,15 °C	0 °R	−459,67 °F	Absolute zero
	255,37 K	−17,778 °C	459,67 °R	0 °F	
	273,15 K	0 °C	491,67 °R	32 °F	Melting point of ice
	273,16 K	0,01 °C	491,69 °R	32,018 °F	Triple point of water
Equal temperature differences	1 K	1 °C	1,8 °R	1,8 °F	
	0,555 56 K	0,555 56 °C	1 °R	1 °F	

Value in °C = $\frac{1}{1{,}8}$ · (value in °F−32)

CONVERSION FACTORS FOR ROUNDWOOD

m3 solid	M3 piled	cord piled	ft3	cunit	MBF board	ft3qg	m3sk
1	1,43	0,394	35,3	0,353	0,211	27,9	1,18
0,700	1	0,276	24,7	0,247	0,148	19,5	0,823
2,53	3,62	1	89,6	0,896	0,538	70,4	2,98
0,0283	0,0404	0,0112	1	0,01	0,00600	0,787	0,0333
2,83	4,04	1,12	100	1	0,600	78,7	3,33
4,72	6,74	1,86	167	1,67	1	131	5,55
0,0360	0,0514	0,0142	1,27	0,0127	0,00763	1	0,0422
0,850	1,21	0,335	30,0	0,300	0,179	23,7	1

BF = board feet, for sawn wood = 1 · 1 · 1/12 feet = 1/12 ft^3. MBF = 10^3 BF. When used for wood in the round = assessed yield of logs, often around 50% of the true volume of logs (as in the table above).

ft^3qg = ft^3 quarter girth or hoppus feet, based on the approximation of the area of a circle to $(\frac{g}{4})^2$ instead of $\frac{g^2}{4 \cdot \pi}$ (g = girth).

m^3sk = In Sweden used unit for the total volume of a bole on bark from the felling cut to the top. Used mainly in inventories of the growing stock and includes all trees that have reached breast height (1,3 m). Approx. equivalent to 0,85 m^3 solid, merchantable wood under bark. Piled wood asumed to have a solid volume of 70%.

SLIP

TO LECTURERS AND TEACHERS AT FORESTRY SCHOOLS

In using "Operational Efficiency in Forestry, Vol 1 Analysis" the mathematical models of forest work operations are understood more easily by the students if explained by numerical examples on the blackboard performed either by the lecturer or one of the students.

As a supplement to the book, a booklet with such numerical examples for the use of the teacher has been prepared. It contains exercises on the applications of the models in numerical examples, including solutions.

For many of the exercises, some points for discussion are also listed, e.g. possible modifications or extensions with additional variables, shortcomings of the model, its validity and its conformity with reality.

The booklet can also be used for preparing cases for written examinations.

ORDER FORM

To FORTEXT, c/o SLU, S-770 73 GARPENBERG, Sweden

Please send ..... copy(-ies) of the booklet "Exercises to Operational Efficiency, Vol. 1 Analysis" by Stig Andersson and Ulf Sundberg to:

Name ....

Address ....

....

....

Payment US$ 10 per copy send (please indicate below):

☐ Cash or

☐ International Money Order in the name of:
FORTEXT, c/o SLU, S-770 73 GARPENBERG, Sweden

The booklet will be sent by air mail without postal charge, including receipt.

MIX
Papier aus verantwortungsvollen Quellen
Paper from responsible sources
FSC® C105338

If you have any concerns about our products,
you can contact us on
ProductSafety@springernature.com

In case Publisher is established outside the EU,
the EU authorized representative is:
Springer Nature Customer Service Center GmbH
Europaplatz 3, 69115 Heidelberg, Germany

Printed by Libri Plureos GmbH
in Hamburg, Germany